WINGS OF DIPLOMACY

MY LIFETIME OF SERVICE
AS AN AIR FORCE ATTACHÉ

COLONEL LEWIS S. WALLACE JR.,
USAF (RETIRED)

WINGS OF DIPLOMACY

Copyright ©2025 by Colonel Lewis S. Wallace Jr., USAF (Retired)

Paperback ISBN: 978-1-965593-72-1

All rights reserved. No part of this publication may be reproduced, distributed, or transmitted in any form or by any means, including photocopying, recording, or other electronic or mechanical methods without the prior written permission of the author except in the case of brief quotations embodied in reviews and certain other non-commercial uses permitted by copyright law.

Published by Cornerstone Publishing

A Division of Cornerstone Creativity Group LLC
Info@thecornerstonepublishers.com
www.thecornerstonepublishers.com

Author's Contact

To book the author to speak at your next event or to order bulk copies of this book, please, use the information below:

erasaya1@yahoo.com

Printed in the United States of America.

"The views expressed are those of the author's, and do not reflect the official guidance or position of the United States Government, the Department of Defense, the United States Air Force, or the United States Space Force."

DEDICATION

I dedicate this book to the legacy of the Tuskegee Airmen, a group of primarily African American pilots and support personnel who served with distinction in the United States Army Air Corps during World War II. Despite facing racism and segregation, they excelled in combat, completing over 15,000 sorties and destroying over 260 enemy aircraft. Their legacy serves as a testament to the power of determination and perseverance.

To the four women in my life. Without their love and support, this book would never have been written.

Maryana Gambrell, Grandmother, Beverly Wallace, Mother, Vivian Wallace, Wife, and Annie Eison Wallace, Wife.

"Pessimists calculate the odds. Optimists believe they can overcome them."

- TED KOPPEL

CONTENTS

ACKNOWLEDGMENTS

I am deeply grateful to my mentor and friend, Dr. Yomi Garrett, whose guidance,encouragment, and support played an essential role in the completion of this memoir.

A LIFETIME OF SERVICE

"To be yourself in a world that is constantly trying to make you something else is the greatest accomplishment."
- Ralph Waldo Emerson

Writing a book takes a significant amount of time, tremendous discipline, and, most importantly, effort on a grand scale. However, books are essential in the final analysis, and creating them is crucial. For the average author, there is a need to decide on a topic, research the various themes surrounding the subject, determine a clear structure for the book, and then actually write it. Writing a book about one's life is an entirely different proposition, primarily because it serves as a memory book. The three primary formats are a biography, an autobiography, and a memoir. A biography is an individual's life history written by someone else. An autobiography is the story of a person's life, written by that person, while a memoir is a collection of memories penned by an individual.

I welcome you to the pages of an *autobiography* and a *memoir.* Because the narrative begins with my early childhood and details the key events of my life, including where I was born and raised, my education, career, life experiences, the challenges I faced, and my key life achievements, it is classified as an autobiography. Naturally, an autobiography is also somewhat subjective. It is written as facts based on one's memories of specific situations.

Such narratives can easily be biased since they reflect one's opinions on particular subjects and account for one's feelings while navigating various life situations.

The word *memoir* is rooted in the French word *mémoire*, which means memory or reminiscence. Similar in many respects to an autobiography, a memoir is the story of a person's life as told by that person. Such life stories often stem from first-person diary entries and accounts. The difference between a memoir and an autobiography is that a memoir focuses on reflection and establishes an emotional connection rather than simply presenting the facts of one's life. Almost invariably, the classic memoir tells a story of great struggle and significant challenges uniquely. The book you hold in your hands combines elements of both a memoir and an autobiography. I have written it to inspire, inform, and enrich the reader.

My principal motivation for writing this book stems from my desire to instill a sense of hope in the underprivileged Black and Brown youth of America. That *message of hope* is that there is a viable alternative to a life of gangs, drugs, crime, and violence. I speak of these youth in the stark light of the crippling conditions that have victimized them, addressing their unemployment and underemployment, severely constrained educational circumstances, abject sense of rejection, and paradoxical yet unconscious self-hatred. These youth have had little or no power over their circumstances and have been helplessly driven to perpetuate their powerlessness. They find themselves in neighborhoods where they are forced to pay more for less and where their resources are relentlessly drained without being replenished. Because of the color of their skin, these youth reside in domestic colonies, leaving them segregated, dominated, and marginalized politically, economically, and educationally.

During the 1950s and 1960s, crimes of theft, extortion, and drugs emerged as desperate means of survival in Black neighborhoods. The police were more concerned with maintaining social control than protecting the poor; this and other factors combined to create the *inner cities that* malevolently imprisoned Black and Brown people. Paul Laurence Dunbar, who lived between 1872 and 1906 and authored numerous collections of poetry and prose, was one of the first African American poets to gain national recognition. He captured the plight of the Black person in these poignant words:

A crust of bread and a corner to sleep in,

A minute to smile and an hour to weep in,

A pint of joy to a peck of trouble,

And never a laugh, but the moans come double; And that is life!

A crust and a corner that love makes precious,

With a smile to warm and the tears to refresh us;

And joy seems sweeter when cares come after, And a moan is the finest of foils for laughter; And that is life!

As I have already stated, life in the ghetto was extremely difficult for Black youth in the 1950s and 1960s. I grew up in abject poverty in the inner city of segregated Washington, DC. The District of Columbia officially became the first majority Black major city in the United States in 1957. Over time, as the Black population grew, it gradually became known as the nation's "Chocolate City," a name popularized by a song rendered by the funk band Parliament, the lead track of their 1975 album of the same name, shortly after the passage of the District of Columbia's Home Rule Act in 1973. This appellation, however, did little to alleviate the pangs of hunger and deprivation prevalent in the inner city. I have always dreamed of a better life where I could escape poverty and become an honest, hardworking, and productive member of society.

I have written my life story in the fervent hope that it will inspire young individuals from underserved and underprivileged communities. It is also my hope that, within the pages of this book, they will find strategies, guidelines, and an overall blueprint that will help them challenge long-held stereotypes about how difficult it is for a person of color to transcend racial prejudice, discrimination, and marginalization, break free of real and perceived barriers, and pursue and achieve their dreams. I hasten to add that I have derived tremendous motivation to write this book from my mother and mentor, Beverly Wallace. On my eighth birthday, I sat across from this dear matriarch for a heartfelt discussion about my future.

"Mum, when I grow up, I want to be the first Negro President of the United States of America," I declared with innocent candor.

If my mother was taken aback by the audacity and precocity of my declaration, she did not show it. Gazing affectionately into my eyes, she held my little hands in hers.

"Son, you can be anything you want in this great country. It just takes believing in yourself, getting a good education, working hard, and never giving up on your dreams."

That singular pronouncement of maternal validation would stand me in good stead over the years. I now realize that the words we speak to our youth are paramount in the eventual outcome of their lives. Many parents and guardians have altered a child's destiny by speaking negatively to him. What most people do not know is that these words are automatically accepted as truth by the powerful subconscious mind, driving the child's self-image into the realm of the extremely low, psychologically incapacitating him and blighting his prospects for advancement for the rest of his life.

With this book, I choose to speak *purpose, potential, and high aspiration into* the lives of Black youth. That is why the target audience of this book is Black and Brown youth across America. In this book, they will learn how a poor Black boy from the inner city of Washington, DC, overcame numerous stereotypes and obstacles to become a federally designated expert on the Russian language and Russian affairs, a colonel in the United States Air Force, and the first Black Assistant Air Attaché in the history of the United States to be assigned to the American Embassy in Moscow.

When all is said and done, the lasting legacy I hope to impart through this book to young people in marginalized communities all over the world is that there is always hope for a brighter future for those who decisively take their destinies into their own hands by declaring that they will refuse to eat from the crumbs falling from the table, but rather take their place at the table and partake of the main course. Finally, I want to use my story to convince you that to be successful in life, you must have unwavering determination, resolute persistence, and stoic perseverance in your quest to achieve your dreams. Happy reading.

Colonel Lewis S. Wallace Jr., USAF (Retired)

Spotsylvania, Virginia, United States of America,

April 2025

CHAPTER ONE

THE EARLY YEARS

"It doesn't matter if you come from the inner city. People who fail in life are people who find lots of excuses. It's never too late for a person to recognize their potential in themselves."
- Ben Carson.

I was born on November 14, 1944, at Gallinger Municipal Hospital in Southeast Washington, DC. Gallinger Hospital primarily served the poor and disadvantaged residents of the city. Located at Independence Avenue and 19th Street, SE, Washington, DC, Gallinger Municipal Hospital was renamed District of Columbia General Hospital in 1953 and closed in May 2001 after serving the residents of the capital territory for nearly 200 years. The controversial closing of this historic healthcare complex in 2001 ended its inpatient services, and the city's indigent healthcare system was primarily transferred to the Greater Southeast Community Hospital. The building currently serves as a homeless family shelter.

My father, Lewis Wallace Senior, was a janitor, while my mother, Beverly, was an office clerk. The first child in the Wallace household was my sister, Barbara. However, because I was the first-born son, my birth was celebrated with expected fanfare by my parents, relatives, and family friends. My maternal grandmother, Mariana Gross Gambrell, a Native American, affectionately nicknamed me "Sonny," and that name has stuck with me ever since. Grandma Gambrell was a profoundly religious and spiritual woman. As a young boy, she constantly read me Bible verses, saying I was a "special child of God" who would accomplish great things. I would fully comprehend Grandma Gambrell's words of unerring prophecy later in life.

In retrospect, I can't help feeling that Grandma's premonition about my future drew inspiration from an essay by W.E.B. Du Bois entitled *"The Talented Tenth."* In his essay, Du Bois argued for the importance of higher education for African Americans and the need for a select group of educated individuals, which he referred to as "The Talented Tenth," to lead and uplift the Black community. To express matters accurately, The Talented Tenth is a term designated as a leadership class of African Americans in the early 20th century. Although the term was created by white Northern philanthropists, it remains primarily associated with W.E.B. Du Bois, who used it as the title of his influential essay that appeared in The Negro Problem, a collection of essays written by leading African Americans and assembled and edited by Booker T. Washington, the educator, author, and orator who, between 1890 and 1915, was arguably the primary leader in the African-American community and of the contemporary Black elite.

When W.E.B. Du Bois wrote *The Talented Tenth in* 1903, Theodore Roosevelt was president of the United States, and industrialization was the order of the day in America. Regarding the Black American population, Du Bois was wary of the

Industrial Revolution, believing it was a propitious time for African Americans to advance their positions in society. His famous essay referred to the one in ten Black men who would cultivate the ability to become leaders of the Black community by acquiring a college education, writing books, and becoming directly involved in social change. In The Talented Tenth, Du Bois argued that these college-educated African American men should sacrifice their interests and use their education to lead and improve the Black community. His firm conviction was that the Black community needed a classical education to reach its full potential, rather than the industrial education promoted by the Atlanta Compromise endorsed by Booker T. Washington, then president of the Tuskegee Institute. In his September 18, 1895, speech to the Cotton States and International Exposition in Atlanta, Georgia, Washington promoted vocational education, industrial occupations, and the learning of practical trades that would provide African Americans opportunities for economic advancement and wealth creation, rather than more intellectually focused pursuits, such as higher education. He urged Blacks to "*cast down your bucket where you are,*" emphasizing his view that they should stay in the South and try to make the most of their situation. W.E.B. Du Bois disagreed with that notion and minced no words about his opposing view in his essay, *The Talented Tenth.* I sincerely believe that I am a product of that dissenting view.

W.E.B. Du Bois, seeing classical education as the pathway to bettering the Black community and as a basis for what, in the 20th century, would be known as public intellectuals, wrote in his essay, *"Men we shall have only as we make manhood the object of the work of the schools; intelligence, broad sympathy, knowledge of the world that was and is, and of the relation of men to it, this is the curriculum of that Higher Education which must underlie true life. On this foundation, we may build bread-winning skills, the skill of hand, and quickness of the brain, with never a fear lest the child and man mistake the means of living for the object of life. "He further wrote, "The Negro race, like all races, is going to be saved by its exceptional men. The problem of education, then, among Negroes must first of all deal with the Talented Tenth; it is the problem of developing the Best of this race that they may guide the Mass away from the contamination and death of the Worst.* "Later, in *Dusk of Dawn,* a collection of his writings, Du Bois redefines this notion, acknowledging contributions by other men. He wrote, *"My panacea of an earlier day was a flight of class from mass through the development of the Talented Tenth; but the power of this aristocracy of talent was to lie in its knowledge and character, not in its wealth."* Today, Black men like me and many others of our (ilk) have W.E.B. Du Bois to thank for a philosophy that guaranteed emancipation from the mediocrity we might otherwise have been eternally stuck with.

During my early years, although we lived in a rundown, rented row house in Washington's inner city, deprived of what, even then, were the conveniences of modern plumbing and running water, and had to make do with an outside toilet, I still had a relatively happy childhood. My parents gave me sufficient love and guidance to soften the sting of poverty. After my mother lost her job as an office clerk, my father refused to allow her to look for another position and, with a finality that brooked no dissent whatsoever, declared that her job was to stay at home and care for the family. I recall the many days when we had no food to eat. However, to help alleviate our hunger and food insecurity, I would often accompany my mother once a month to the Welfare Surplus Food Distribution Center in my red wagon and pick up canned fruits and vegetables, cheese, meat, and other staples. Mother was such an excellent cook that she could be counted on to prepare wholesome dishes from scratch.

I began elementary school at an all-Black school in my neighborhood when I was five. I was an above-average student and excelled in mathematics, science, and foreign languages. My favorite subject was French, and I always received excellent grades. My parents bought me a chemistry set for my eighth birthday to encourage my interest in science. However, a near-tragic incident, in which I started a fire that almost burned down my bedroom after mixing various chemicals at home, signaled the end of my interest in chemistry. My experiences in inner-city schools, like the one I attended, varied. Children came to school with different backgrounds that affected their performance. Some children arrived ready to learn, while others did not. The problem with inner-city elementary education likely lies in not having the means to work with all students in a way that meets their needs, even if it means first addressing their basic life needs of safety, food security, and adequate sleep. In my opinion, the disadvantage of educational marginalization has become a cultural problem that glaringly highlights the stark differences between education for the underserved and education for those in ideal learning circumstances. Without a doubt, in my early school days, education for inner-city school children differed significantly from that for children in more affluent neighborhoods. It was clear that school segregation eloquently described the social divide. Socioeconomic disadvantage affected the education that inner-city school children received. Yet, all the inner-city children needed were the right tools. Against all odds, however, I seemed to flourish at school, and it was clear to everyone that I was poised for a future that would be a marked departure from what fate might ordinarily have had in store for me.

My parents were very religious. Every Sunday, it was traditional, if not mandatory, for the family to attend church service. I recall spending the entire day in church, from Sunday School to Church Service and onto Baptist Training Union sessions. Since our home church was in the Georgetown section of Washington, we had to walk approximately three miles one way to attend. At that age, my comprehension of the Bible was still in its infancy, and I relied heavily on my parents' knowledge to fill the gaps. Yet, as a poor youth from the inner city of Washington, DC, my early religious upbringing gave me a strong sense of belonging, moral guidance, mentorship, a genuine love for education, and personal growth. It also helped me develop resilience in the face of the adversity I would inevitably encounter later in life. Growing up in my days was no less complicated than growing up in today's world. Parents, especially in the Black community, were often worried about how their children would navigate the social, behavioral, and developmental challenges of a new age of permissiveness, especially during adolescence, which can be a challenging time. My religious upbringing profoundly helped me navigate those challenges.

Against the background of this seemingly Puritan upbringing, my first encounter with law enforcement came as a surprise, if not a shock. In astonishing variance with my religious and moral upbringing, my first encounter with the Washington, DC, Police Department occurred when I was just ten years old. My sister, Barbara, and I were so deprived that we had little or no money to buy the toys and candy that are the usual preoccupations of children. To solve that challenge, we devised a somewhat dubious scheme. We decided to shoplift at a local convenience store. For our relatively tender ages, we were quite creative in our planning. We devised a plan to cut holes in the pockets of our jackets so that the stolen goods would fit inside the clothing liners. Unfortunately, luck was not on our side. The store clerk stopped us as soon as we attempted to leave, and we were apprehended for the attempted theft of merchandise. The store manager called the police and notified our parents of the incident. Fortunately, the store manager refused to press charges, and the police released us into the custody of a disappointed father. To this day, I still recoil at the sheer vehemence of our parents' scolding and the severity of the punishment that followed.

CHAPTER TWO

GROWING UP IN THE "HOOD"

"It doesn't matter where you're from. You can be from the hood or the suburbs. You can be poor with no education or a college graduate. No matter your background, you can win."
- **Brandon Christopher McCartney,** more famously known as Lil B, the rapper.

While the prevailing ills of segregation and poverty combined to impose numerous daunting challenges and severe limitations on the Black residents of the inner city of Washington, DC, in the 1950s and 1960s, there were *positive aspects* to those seemingly disadvantaged circumstances, as well as *negative aspects* to living in the city during my childhood. First, permit me to delve briefly into the history of Washington, DC, so that my narrative can be easily understood from a proper perspective.

The history of Washington, DC, is tied to its role as the federal capital of the United States.

President George Washington first selected the site of the District of Columbia along the Potomac River. Unique among American cities with a high percentage of African Americans, Washington, DC, has had a significant Black population since the city's creation. As a result, the city became both a center of African American culture and a hub of the Civil Rights Movement. The United States capital was initially located in Philadelphia. In September of 1791, combining the topographical name of *Columbia* and the last name of the sitting president of the United States, *Washington,* it was agreed to name the federal district as the T*erritory of Columbia and* the federal city as the *City of Washington.*

Throughout its history, Washington, DC, has undergone massive transformations.

Starting as a sparsely populated plantation society that grew to be a center of the slave trade, America's seat of government developed into a diverse metropolis over time, eventually becoming the nation's first majority Black city. Over the years, there have been tremendous tensions between race and democracy, along with significant milestones in the evolution of segregation, gentrification, and civil rights in Washington, DC. The 1950s and 1960s were a particularly intense period in the city's history. The Black urban population experienced riots following the assassination of Martin Luther King Jr. They were struggling to build a new society with less racial discrimination while primarily residing in the inner city. High rates of vagrancy, poverty, and crime were prevalent due to a general lack of education and stable employment. Yet, it must be noted that segregation in Washington, DC, was a glaring example of the contradictions in American society. In the 1950s and 1960s, the city's government, including schools, was under the control of Congress. Its members proudly portrayed the city as the capital of the free world, where democracy and personal freedoms were defended against the threat of communism. However, most of the city's public facilities, schools, and housing were segregated by law or practice. The phrase "inner city" *gradually* became widely used to refer to the city's older, poorer, and more densely populated central section. Across the United States, inner cities are distressed urban and suburban areas of concentrated poverty and low income, with inner city residents representing 14% of the population, about 45 million. In the final analysis, of all American residents living in poverty, 31 percent reside in inner cities.

On a *positive* note, however, despite the many restrictions imposed by segregation, the residents of my inner-city community, albeit abjectly impoverished, continued to strive culturally and economically. As a young boy growing up in Washington, DC, I nostalgically recall an ethos of cultural vibrancy in the city. For example, the U Street NW corridor was a throbbing hub of black-owned businesses and artistic activity. In fact, in the vicinity of my neighborhood, there were numerous black-owned convenience stores, banks, movie theaters, restaurants, barber shops, florists, and doctors' offices. To enrich a burgeoning and vibrant economy, the creative works of Black artists, musicians, and intellectuals combined to provide a rich cultural experience for the city's residents. Indeed, I grew up in Washington, DC, and witnessed an intellectual and cultural revival of African American music, dance, art, fashion, and literature. For instance, Duke Ellington, an internationally acclaimed composer and musician, was a native of Washington, DC, and played a significant role in the Black community during that era. Naturally, other cultural expressions were inspired by a renewed militancy in the general struggle for civil rights, combined with the migration of African American workers fleeing the racist conditions of the American South.

It is virtually impossible to discuss any period in the history of Washington, DC, without referencing the contributions of Black Americans. Since the city's inception, Black culture has been entirely intertwined with the nation's capital, making it the ideal place to learn about Black history. This is because, even today, Washington, DC remains a cornerstone of Black history as the site of Howard University, one of the most well-known Historically Black Colleges and Universities, and historic Black-owned restaurants that once served as meeting venues for leaders of the Civil Rights Movement. U Street, once widely known as *Black* Broadway, is representative of that era in the city's history but has now become a prime example of ongoing gentrification. U Street was the heart of the Black community during segregation.

There was also a tremendous sense of community among Black residents of Washington, DC. Churches, schools, and social organizations, such as the Boy Scouts of America, the Metropolitan Police Boys Club, and the YMCA, all provided moral and mentoring support to help me navigate the severe difficulties of daily life in the inner city. Although schools in my neighborhood were extremely underfunded in terms of new textbooks, supplies, and crumbling infrastructure, I distinctly remember a dedicated group of teachers and administrators who worked to provide quality education for their students. In addition to my academic accomplishments, I excelled in sports, including baseball and basketball, as well as acting and choir.

In the face of the harsh realities of poverty and segregation in Washington, DC, my childhood, as if to serve as a soothing balm for those brutally austere days, was spiced with joyful interludes with my family. My mother, a graduate of Dunbar High School in Northwest Washington, DC, always emphasized the importance of a well-rounded education. To supplement our traditional public school curricula, my mother would take the family on excursions to the many free museums and art galleries of the Smithsonian Institution in Downtown Washington, DC, on weekends. It was very convenient that Washington, DC, boasted free Smithsonian museums. Collectively called The Smithsonian Institution, this world-renowned museum and research complex in the District of Columbia, consists of seventeen museums, galleries, and a zoo. Each one is free to enter, and across the spectrum, one can learn about human origins, art's wonders, and so much more. The free entrance to the museums sets the Smithsonian apart from other world-class cultural institutions. It allows millions of visitors from all walks of life to experience and learn about science, art, and history. The Smithsonian, being only partially funded by the federal government, relies on financial support from private individuals, corporations, and foundations to accomplish its mission to *"increase and diffuse knowledge"* and to preserve America's cultural heritage for future generations. Without a doubt, the generosity of these individuals and organizations demonstrates their love for the country's history and appreciation of the Smithsonian's role in preserving and sharing that history with others. I remain grateful that, as a family, we could take advantage of the unique opportunity to learn that the facility provided us. Naturally, I am even more thankful to my mother for having the cultural awareness to expose us to this chance.

After our weekly excursions, we often stop at the Woolworth's Department Store on 14th Street in NW Washington, DC, for lunch. Unfortunately, during the 1960s, Woolworth's was known for its segregationist policies, which prevented Black people from sitting at the lunch counter alongside white customers. Therefore, we were forced to go to the rear counter of the restaurant to order and pick up our food. Additionally, every year during the Easter holidays, the family would visit the Washington DC Zoo to participate in the annual Easter egg hunt and picnic. Regretfully, although the White House also had its own unique and historic yearly Easter egg hunt, Black people were prohibited from attending it during my formative years. During the summer months, my father would load the entire family into the back of his old work truck to visit one of the Black-only beaches in nearby Maryland. Carr's Beach in Annapolis and Sparrow's Beach in Anne Arundel County served as essential gathering places for African American families, providing a sense of community and leisure during racial segregation. Carr's Beach, in particular, was a highly popular Chesapeake Bay resort and concert venue for the African American community during the years of segregation. Unfortunately, Carr's Beach is no longer a beach destination, but its legacy lives on. Affectionately called "the Beach," Carr's Beach hosted many major African American artists performing for Black audiences during the 1940s, 1950s, and 1960s. Located on the Annapolis Neck, just south of Annapolis, Carr's Beach was one of several private bayside beaches available when pervasive segregation denied African Americans access to white-only beaches and facilities. In this intense era of forced racial segregation, Black audiences and artists responded out of necessity by creating their own recreational attractions and performance venues. As a result, Carr's Beach gained prominence as a remarkably popular and successful African American attraction.

The negative aspects of life in the inner city, or more bluntly, the *"hood,"* are quite poignant. The *hood* is an abbreviation of "neighborhood." In many respects, it is not necessarily synonymous with ghetto; it refers more to *disadvantaged urban culture.* Although the phrase *"in the hood"* can be used more broadly to refer to being in a particular community or social environment, especially one characterized by a specific subculture or way of life, it more commonly refers to being in or from a neighborhood perceived to be disadvantaged, often characterized by poverty, crime, or other social challenges. Living in the hood is traumatic. It involves witnessing a teenage boy shot down in front of your apartment. That teenager might only have wanted to protect his sister from her drunk boyfriend. Living in the hood means watching other Black boys try to save him, knowing he is already dead. It is seeing fifty police officers among many witnesses to the crime, yet not everyone knows what happened. It is watching his sister wail as she and her father receive the gruesome news that the fifteen-year-old boy is no longer living. Living in the hood is seeing that fifteen-year-old boy's memorial service turn into an all-out fistfight between thirty teenagers. Living in the hood means that by the next day, everything has returned to normal, and there is a ribbon around the tree next to where the poor boy died.

Living in the hood is having your best friend's five-year-old son shot in the head while eating cereal at the kitchen table. That five-year-old was collateral damage, as his father was the target. He owed money but couldn't pay it back. Living in the hood is a drunk drug dealer slamming into the back of a school bus and nearly killing your little brother. Living in the hood is your father right outside in the middle of the road, engaged in a fistfight because your mother's best friend was smacked in the face for asking a man "*doing crack*" in the hallway to leave before the children came out and saw him *"doing crack."* Living in the hood means merely *scraping by,* never knowing if you will eat that day, but doing everything in your power to ensure your children eat three square meals daily, no matter what it takes. It means living from *paycheck to paycheck*. Living in the hood means clutching your phone to your chest as you walk to work so it doesn't get stolen. Living in the hood is facing all those uncertainties and still trying your hardest to raise sane, healthy, and productive children. Living in the hood is knowing that all these are merely the tip of the iceberg of the experience of living in the hood.

Living in the *"hood"* affected my daily existence and ultimate survival. For example, most of the time, my father could only find seasonal construction work. As a tragic consequence, there were never sufficient funds to pay the bills or buy food. In just five years, we were compelled to move home four times due to failure to pay the rent and utility bills. At the tender age of ten, my sister, Barbara, and I became unwitting participants in the Civil Rights Movement in Washington, DC. My family had moved from Northwest Washington to a dilapidated row house on Lincoln Road in Northeast Washington. As I mentioned earlier, segregation was a significant issue in schools in the 1960s, and the nearest school to our home was Emory Elementary, an all-white school. When Emory was forced to desegregate in 1954, Barbara and I became some of the few Black students to attend the school. I recall how White parents protested vigorously every day in front of the school because Black students were enrolled in classes with their children. It was a terrifying time for my sister and me, and we had to have a police escort accompany us for the few blocks to and from school.

When I was eleven, we moved to the LeDroit Park neighborhood of Washington, DC. This was a turbulent time when residents of LeDroit Park faced severe challenges with racial tension, crime, poverty, and deteriorating infrastructure. Additionally, there were constant flashes of gang violence in the areas surrounding LeDroit Park. At the tender age of twelve, I was forced to become a member of the notorious *'LeDroit Park Crew"* or suffer the consequences of frequent beatings by gang members. The LeDroit Park Crew operated in the neighborhood and surrounding areas, gaining a fearsome reputation for their involvement in street crime, including drug trafficking, theft, and acts of violence against rival gangs and individuals. Ironically, however, despite their notoriety, the LeDroit Park Crew also provided a sense of community and protection for some residents in the neighborhood. My mother didn't fully comprehend the true nature of the LeDroit Park Crew, believing it was just a neighborhood social club. She even proudly purchased me a sweater with the gang's logo. Needless to say, I never confided in my mother about the organization's true nature.

This phase of my youth was also my Junior High School years. It was a time when my life was full of stark contradictions. On one hand, I strived to be a good student and even participated in the noble and uplifting activities of youth social clubs such as the Boy Scouts, the Boys Club, and the YMCA. *On the other hand, under the murky cover of darkness and shadows, I used the knives, hatchets, and other* weapons I had acquired from the Boy Scouts to engage in combat, otherwise called rumbles, with warring street gangs. I consider myself fortunate to have come out of gang life alive. In any case, my life certainly wouldn't have taken the trajectory it later did if I had remained a gang member, dead or alive. Thankfully, my association with the LeDroit Park Crew ended after about a year, when the gang began participating in numerous armed robberies and assaults with intent to kill. I sincerely submit that I am not at all proud of my tenure as a "gangbanger." Yet, as a young boy growing up in the inner city in the 1960s, gang membership was the only way for me to survive. In addition to my sincere desire to reject a life of crime, another motivation for my decision to turn my life around came from a longtime friend and fellow gang member named Moe. Moe had been sent to a juvenile detention center, known as a reformatory, on several occasions for petty crimes. One day, after his release from detention, Moe and I had an earnest conversation that would greatly impact my future. He pleaded with me to leave the gang and turn my life around so I could accomplish great things. Today's story is nothing short of my decision to heed that advice. I am glad that I did.

To close this chapter, I will offer the possible reasons why someone might join a gang. The first reason is *poverty*. Most gangs exist primarily as moneymaking enterprises. By committing theft and dealing drugs, gang members can make relatively large sums of money. People who lack money may turn to crime if they can't earn enough from a legitimate job. This partly explains why gangs exist in America's inner cities. However, I must add that not everyone who is poor joins a gang, and not every gang member is poor. The second reason is *peer pressure,* which was a factor in my membership in the LeDroit Park Crew. I felt compelled to join the gang to protect myself from frequent beatings by other gang members.

Additionally, there is the factor that gang members tend to be young. This is partly because gangs intentionally recruit teenagers and because young people are very susceptible to peer pressure. If they live in a gang-dominated area or attend a school with a strong gang presence, they may find that many of their friends are joining gangs. It can be difficult for a teenager to understand the harm that joining a gang can cause if they're worried about losing all their friends. Many teenagers resist the temptation of gang membership, but for others, it is easier to follow the crowd. The third reason is boredom. With little else to occupy their time, youths sometimes turn to mischief for entertainment. If gangs are already present in the neighborhood, that can quickly provide an outlet. Alternatively, teenagers might form gangs themselves. This is why many communities have tried to combat gangs by simply giving kids something to do. Dances, sports tournaments, and other youth outreach programs can keep youths off the streets. Indeed, youth sports programs, or even libraries, are often in short supply in inner cities. Thankfully, for every teenager who gets bored and joins a gang, there are ten others who find positive, productive ways to spend their time. The final reason is despair. If poverty is a condition, despair is a state of mind. People who have always lived in poverty, with parents who lived in poverty, often see no chance of ever getting a decent job, leaving their poor neighborhood, or obtaining an education. Drugs and gangs surround them, and their parents may even have an addiction themselves. A neighborhood gang can seem like the only real family they will ever have. Joining a gang gives them a sense of belonging and a part of something important that they can't find elsewhere. In some cases, parents approve of their children joining gangs and may have been members of the same gang in the past.

Drug use is an underlying factor in all of these reasons. Not only does the sale of illegal drugs drive the profits of street gangs, but it also creates many of the conditions that lead to gang membership. The brutal truth of gang life is that the only way most gang members leave is in a *body bag*. Some do manage to move on to a better, peaceful life, as I was fortunate enough to do. That might be because they attain a level of maturity that allows them to see the dangers of gang life from a different perspective. They want to protect their loved ones if they have a family or seek a good job and a home. Ultimately, there is no easy way to stop gangs because the underlying conditions that lead to gang formation are complex. Police crackdowns can temporarily lower gang influence in a specific area. However, when poverty and despair remain, gangs will inevitably recruit new members to replace those who are incarcerated. In the final analysis, therefore, the fate of a young Black man concerning gang membership ultimately rests on strong personal initiative, as was the case with me.

CHAPTER THREE

HIGH SCHOOL DAYS: JOY AND SORROW

"A high school diploma will no longer be sufficient. But that post-secondary education does not have to be a four-year university or a four-year college. It can be career technical education, vocational education, or community college."
- Raja Krishnamoorthi

I quit the LeDroit Park Gang when it became clear that its members had graduated to the status of armed robbers and were willing to engage in physical assault to commit murder. Although this realization did not necessarily operate at my subconscious level, especially given my relative youth at the time, it is a well-acknowledged fact that replacing an unwholesome behavior with an alternate behavior is a good way to change that behavior. It was clear that I had decided to make a positive change in my life. For me, quitting the LeDroit Park Gang also meant an immediate commitment to *two* strategies for successfully implementing that decision. *First,* I focused on my academic studies. *Second,* I devoted myself to positive community activities. In particular, for combined moral and mentoring support during that critical period, I turned to church, school, and social organizations, such as the Boy Scouts of America, the Metropolitan Police Boys Club, and the YMCA.

In junior high school, I performed exceptionally well in Science, English, and Latin, and I graduated in the top ten percent of my junior high school class in 1960. We lived in the Upper Northwest section of Washington, DC, at the time of my graduation. Although during that era, educational differences and inequalities between schools in privileged and predominantly Black neighborhoods created a sense of urgency for very low-income parents, particularly parents of African American students, my parents were understandably unwilling to see me attend a high school in our predominantly Black neighborhood. One couldn't blame my parents for those sentiments. A pernicious combination of social and economic disadvantage, all wrapped up as systemic poverty and a host of associated conditions, served to depress student performance in the segregated schools of the 1950s and 1960s. Concentrating students with these disadvantages in racially and economically segregated schools only worsened the situation.

It was a vicious cycle. Schools that the most disadvantaged Black children attended were segregated because they were located in segregated, high-poverty neighborhoods, far from genuinely middle-class neighborhoods. Living in such high-poverty neighborhoods for multiple generations created a barrier to achievement, with multigenerational segregated poverty characterizing many African American families to this day. Education policy was inevitably constrained by housing policy, especially in those days. Indeed, it was not possible to desegregate schools without also desegregating both low-income and affluent neighborhoods. Yet, the policy motivation to desegregate neighborhoods was hindered by a deliberate ignorance of America's racial history. Thus, it became convenient for policymakers to assert that the residential isolation of low-income Black children was simply an *accident of economic circumstances.* The facts, however, were more glaring. Historical records showed that residential segregation resulted from racially motivated public policy, and its effects endure to the present. I mention this because, without awareness of this aspect of our history, policymakers are unlikely to take meaningful steps to understand or fulfill the compelling mandate to remedy the racial isolation of neighborhoods or the schooling disadvantages that flow from it.

Fortunately for my parents, my high standardized test scores and overall academic achievement in junior high school would serve them well as they petitioned the school board to allow me to attend a high school outside my designated zone. During the 1960s, the DC Public Schools (DCPS) instituted a school zone system through which students were assigned to specific schools based on their residential address. To express matters with stark clarity, DC law required each child to be assigned to a school determined by lines drawn around the area where the child lived. That school was called the *"boundary" or "by-right" school.* Students had the right to attend the boundary school if they lived within the designated area. More elaborate patterns determined which high school students had the right to participate based on the *school* they were coming from.

That school zone system played a significant role in perpetuating educational inequality, as schools in predominantly Black neighborhoods were severely underfunded and, as I previously mentioned, lacked resources compared to schools in predominantly white neighborhoods. I was finally selected to attend McKinley Technical High School in Northeast

Washington, DC. McKinley Technical High School was a prominent vocational and technical high school known for its strong academic programs in various technical fields. Located at 2nd and T Streets NE, in the Elkington area of the District of Columbia, the school was named for William McKinley, the 25th President of the United States. Initially, the school was exclusively for white residents of the City of Washington. However, it was integrated with other DC schools by an executive order from President Dwight Eisenhower in June of 1954. Subsequently, the school underwent a rapid transformation in ethnic attendance, similar to other schools in Washington, DC, and by 1960, it had become a majority African American school.

I was well-suited to what McKinley Technical had to offer. The school provided a wide range of courses in engineering, electronics, automotive technology, medical technology, and other vocational trades. I enrolled in the Pre-Engineering Program, and my academic curriculum included courses in Architectural Design, Electronics, Aviation, Mathematics, History, English, and Latin. Of great significance, it was at McKinley that I first discovered my latent passion for flying. We had a *Link Trainer,* otherwise known as a simulator, in the classroom, and in what might have been a premonitory template for the grand scheme of things in my later life, I learned the basic principles of flight right from my high school days.

My introduction to the basic principles of flight in high school seamlessly tied in with my early initiation into military life, and at this point in my narrative, I consider it crucial to highlight my first experience with the military while I was still in high school. In the 1960s, the Army Junior Reserve Officers' Training Corps (JROTC) program was established in high schools across the United States, including Washington, DC. The JROTC program aimed to instill in high school students the noble values of citizenship, service to the community, and leadership. My school, McKinley Tech, had a robust Army JROTC program, and all male students in grades 10-12 were required to enroll. In addition to classes on leadership, citizenship, and military history, students participated in drills, military ceremonies, and physical fitness training.

My participation in the JROTC program was my first experience of what it would be like to be in the military. Notably, my high school years coincided with America's controversial involvement in the Vietnam War. During that time in the 1960s, the United States became mired in an expanding conflict in Vietnam. It would ultimately turn out to be America's longest and most controversial war, with around 2.5 million American men and women serving in Vietnam and over 58,000 losing their lives. This American intervention, which finally ended in 1973, cost billions of dollars and cost Lyndon B. Johnson the presidency. It fractured the national consensus on foreign policy, eroded morale within the military, and sparked massive protests and violence in the United States. In the end, South Vietnam fell to the North Vietnamese Communists in 1975. Yet, for all its controversy and tragedy, the Vietnam War seemed the logical, if problematic, course of action for American policymakers.

Despite the ever-present howling of the winds of war around me at that time, I had no interest in joining the military after graduation. Additionally, it would turn out to be a matter of sober regret that my failure to devote sufficient seriousness and diligence to the JROTC program left me with the lowly rank of *"Private"* for the entire duration of my three-year tenure in high school. My lack of enthusiasm for the military confounded and disappointed my beloved mother. Indeed, it was a source of profound sadness for her that, in my senior year, I did not become one of the cadet officers who wore a distinctive uniform and carried a saber. Yet, with the wisdom of hindsight, my time in JROTC positively influenced my evolution into the man I eventually became. The character-building, leadership skills, and discipline I learned as a member of the high school JROTC program clearly formed the basis of my decision to join the United States Air Force later and serve my country for over thirty years as an officer, aviator, and diplomat.

In addition to my academic studies at McKinley, I excelled at sports, especially basketball. Being an accomplished basketball player in high school was key to my inclusion in many social cliques on campus. As a youth growing up in Washington, DC, I was also exposed to *dancing,* which was an essential part of the social life of Black teenagers. I was such a versatile and gifted dancer that my friends began referring to my dancing as *"poetry in motion"* on the dance floor. Soon enough, I was selected as a "regular dancer" on the local Teenarama Dance television program. This all-Black show was a cultural phenomenon showcasing trends and styles among teenagers during the 1960s. The Teenarama story itself was quite fascinating. Several metropolitan cities in the United States produced local television dance shows in the 1960s, including Philadelphia's "American Bandstand" and "The Buddy Dean Show."

"Show" in Baltimore, Allen Freed's "Big Beat" in New York City, and "The Milt Grant Show" in DC. Although those shows had a strong following, they were lacking Black teens in their earlier broadcasts. In what would turn out to be a significant development for the Black community in 1963, DC's WOOK-

TV launched "Teenarama," the first Black dance program in the country, on the first television station with programming exclusively for Black audiences.

The history of this landmark television show was later captured in the 2006 Emmy Award-winning documentary, *"Dance Party: The Teenarama Story."*

"Teenarama" aired live from Monday to Saturday and was produced from studios that housed both WOOK Television and Radio on First Street NE, a block from the present-day Ft. Totten Metro Station. We arrived after school, well-dressed for dancing and to enjoy lip-synced performances from top national music acts. In addition to showcasing my contemporary dancing skills on television and at numerous *"house parties,"* I also enhanced my social graces by learning how to waltz. Waltz, a dance in which two dancers move in triple time as they turn together in circles, is also the name of the music written for that sort of dance. When you waltz, you face your dance partner with one hand on their waist or shoulder and the other clasping their hand, with the two of you moving in a gliding, almost dizzying spin. The world's most famous *waltz, "On the Beautiful Blue Danube," by Johann Strauss, Jr., celebrates the city of Vienna and the river that runs through it. I would go to the local YMCA for dance classes twice a month. Since there was a shortage of young Black teenagers at Mckinley Technical who knew how to waltz, and although I was poor and lived in the hood, several "upper-class"* young ladies were compelled to invite me to serve as their social escort to debutante balls, cotillions, and proms.

Despite my good times and positive experiences in high school, this period was also marked by great sorrow. During my sophomore year, I was diagnosed with a serious illness. It all started with a ringing sensation in my ears, accompanied by seizures that came with a somewhat confusing unpredictability. The doctors diagnosed my condition as epilepsy. Because of this condition, I remained bedridden for five months, completely unable to attend my classes in person. Instead, my mother arranged for the school to send all my coursework to our home. Through sheer fortitude, determination, and perseverance, I completed all the assignments and advanced to junior year with my classmates. While confined to home, I also found tremendous enjoyment in reading many works of literature considered the classics. My favorite works included *"Moby Dick," "To Kill a Mockingbird," "War and Peace," "The Odyssey," "The Adventures of Huckleberry Finn," "Treasure Island," "The Three Musketeers,"* and *"Julius Caesar."* After consulting with various neurologists in the Washington, DC, area, my mother finally located a neurologist at Howard University Hospital who, after thoroughly researching my case, placed me on a series of antiepileptic medications to control my seizures. Miraculously, I was cured of epilepsy and never had another seizure in my lifetime. After overcoming this illness, I joined my classmates for in-person classes in my junior year and graduated from high school with my class in 1962.

To conclude the narrative of this chapter of my life, I will express a mild opinion on the dislocations caused by segregation. I am a living witness to how our nation's public education system has not served students equitably, especially students of color and those from low-income backgrounds. The consistent storyline that characterizes our system is that schools predominantly serving Black and Brown students tend to have the least resources and are the most underfunded. Even as a young student, the funding and resource inequities across schools and neighborhoods were painfully apparent. It was clear to me that, in most cases, funding and educational opportunities were scarce if your school was just a traditional neighborhood institution located in a poor area. Inadequate public education funding has been a perpetual and devastating problem, particularly for students of color and those in areas of concentrated poverty.

In my opinion, addressing inequality demands serious and, admittedly, uncomfortable conversations about unresolved biases related to race and class. America cannot afford to ignore the big, complex picture of inequity in education. We must resist the urge to oversimplify our issues and commit ourselves to an integrated domestic policy that can prepare children—regardless of their background—for their futures. Equity must start at the top by implementing federal policies that support teachers and students. This means the government must fully address inequality by funding schools in disadvantaged communities and prioritizing youth. We will receive only what we invest, not just for our schools but for our entire society and for the democratic foundation that guarantees the right to equal opportunity for all citizens.

Perhaps another way to address inequality is through a community school concept. These will be public schools that partner with families and local organizations to support young people's full development and growth. These schools will provide a level playing field for learning because each community school will reflect local needs, assets, and priorities. No two schools will look exactly alike. Such community schools will demonstrate how all schools, policies, and future programs must be coordinated in a comprehensive, adaptive, and, above all else, equitable manner. At a time when inequality in our education system remains dismally high, we all must work together to fund our schools and not be deterred by the challenges ahead. Instead, let us forge ahead and recommit ourselves to the promising potential of a public education system that both government and community fully support and that provides equitable opportunities for all students. Ultimately, however, we cannot substantially improve the performance of the poorest African American students—the "truly disadvantaged," as it were—through school reform alone. This must be addressed primarily by improving the social and economic conditions that leave too many children unprepared to take advantage of what even the best schools offer.

CHAPTER FOUR

UNCLE SAM COMES CALLING

"I Want You!"
- Uncle Sam

After graduating from high school at the age of seventeen, I didn't have the slightest clue what the immediate future held in store for me. Many of my McKinley Tech classmates had already received acceptance letters from various colleges and universities nationwide. Some of my closest friends planned to attend Historically Black Colleges and Universities (HBCUs), such as Howard University, Hampton University, Morgan State University, Tuskegee University, Morehouse College, and North Carolina A&T State University. HBCUs are higher education institutions in the United States, established before the *Civil Rights Act of 1964* to serve the African American community primarily. Today, HBCUs continue to serve the African American community, though most have more ethnically diverse populations than in past generations. Nevertheless, HBCUs are a source of accomplishment and great pride for the African American community. They offer all students an opportunity to develop their skills and talents regardless of race. These institutions train young people who go on to serve domestically and internationally in the professions, as entrepreneurs, and in the public and private sectors.

Given how popular HBCUs had become among promising Black students, that should have been the logical path for me to take. However, two factors immediately precluded that option. *Firstly,* since no one in my family had ever attended college, the thought of earning a degree was as remote from my mind as the Siberian Desert. The very idea never crossed my mind. *Secondly,* my family's dire financial situation was such that the notion they could afford my college tuition was one I needn't even bother to entertain. Because of these twin considerations, I did not apply for college scholarships, even though I scored high marks on the Scholastic Aptitude Tests (SAT).

At that point, it did appear as if I was so directionless that I had no idea what I would do with the rest of my life. Perhaps the only saving grace was that my severely, yet understandably, constrained mind still saw a possibility for survival and the hope of leading a rewarding life by simply finding a decent job in the Washington, DC, area. Yet, in the 1960s, employment options for young Black male high school graduates like me were so limited that they could only be described as few and far between. Not surprisingly, many Black males faced near-insurmountable barriers to finding meaningful employment, let alone securing well-paying jobs. In these circumstances, they often had to settle for low-wage, manual labor positions.

In the summer of the year I graduated from high school, I was fortunate to secure employment as a stock clerk and delivery driver at a small grocery store in the Georgetown section of

Northwest Washington. The salary was one dollar per hour, with the potential for tips. Although this job marked the beginning of my employment journey, I still felt the urge to do something infinitely better with my life. In August of 1962, I decided to submit my application for enrollment in the Freshman Class at the District of Columbia Teachers College (DCTC). DC Teachers College was established in 1955 as a public college in Washington, DC, and was part of the DC Public Schools System. The merger of two respected teachers' colleges, Miner Teachers College and Wilson Teachers College, aimed to prepare teachers for the city's school system. Before being admitted to DC Teachers, all applicants were required to pass a comprehensive entrance examination. In the year of my application, one hundred and fifty applicants took the entrance examination, but only fifty were approved for admission. Thankfully, I was one of the successful applicants and began my studies in September of 1962. I had always been drawn to foreign languages, and because of that love, I decided to major in Spanish with a minor in Science. Although the first two years of college were rigorous and demanding, I excelled in all my courses. However, my life would inadvertently take another direction in 1964. As might be expected, I needed to earn income as a student, so I applied for and was hired by the US Department of State as an entry-level Diplomatic Pouch Clerk in the State Department's Courier Operations Division.

On the Sunday before starting this new job, the Department of State invited the new employees and their families to attend a series of briefings and guided tours of the facility. My mother and I graciously accepted this invitation. We took a taxi to the Main State Building at 2201 C Street, NW Washington. After a series of briefings by State Department officials, we were taken on a building tour. The tour's highlight came when we took the elevator to the eighth floor to view the elegant Diplomatic Reception Rooms. These rooms are the all-important venue for high-level diplomatic events, official receptions, and important meetings hosted by the United States Secretary of State and other senior officials. Our guide told us about the history and significance of these rooms, as well as the role they play in US diplomacy.

To underscore the history and significance of the Department of State's Diplomatic Reception Rooms, our tour guide informed us that they are one of the most extraordinary reflections of America's remarkable cultural accomplishments in fine and decorative arts of the 18th and 19th centuries. Here, visiting heads of state, heads of government, foreign ministers, and other distinguished foreign and American guests are officially entertained. The Diplomatic Reception Rooms have been the gracious setting for many of America's most significant diplomatic achievements. Characterized by their rare beauty, classical balance, and quiet dignity, these State Rooms convey a vivid American spirit. Indeed, the Diplomatic Reception Rooms are home to an unparalleled collection of fine and decorative art from the time of America's founding and formative years, and, as we were told, they are all gifts of American citizens to the nation. These objects come together to reflect the American cultural heritage that the American people proudly share as their ambassador to the world. After the tour, I told my mother I dreamed of someday attending diplomatic receptions in these magnificent and well-appointed rooms. Words, as they say, embody tremendous power, and my dream would later turn out to be a self-fulfilling prophecy when, later in life, I found myself a proud invitee to official receptions hosted by two Secretaries of State and the Secretary of Defense in the dignified ambiance of the Diplomatic Reception Rooms.

I started work as a Diplomatic Pouch Clerk at the State Department, and since my office hours were usually from 4 pm to midnight, I could devote time and attention to my studies during the daytime. After almost a year at the job, I decided to decrease my college load to part-time so I could work full-time at my government job. In hindsight, this was a massive mistake on my part. What happened was as unforeseen as it was life-altering, taking me along an entirely unanticipated trajectory. Shortly after I became a part-time student, Uncle Sam came calling. *Uncle Sam,* which shares the same initials as the *United States,* is a common personification of the federal government. Since the early 19th century, Uncle Sam has been a popular symbol of the US government in American culture and a manifestation of patriotic emotion. Nearly everyone is familiar with the army recruiting poster in which Uncle Sam points directly at the viewer, declaring, *"I Want You!"* That declaration did not mince words in its unequivocal call to patriotism. I received my draft notice to report to the Army Draftee Processing Center at Fort Holabird, located a few miles southeast of Baltimore on the Patapsco River in Maryland. This facility served as a reception and processing center for incoming draftees before they were assigned to basic training camps or specialized training programs. All draftees underwent medical examinations, received vaccinations, and completed a series of aptitude tests and other administrative paperwork before being dispatched to their respective units, primarily in Vietnam at that time.

During the 1960s, the *Selective Service System* in the United States disproportionately impacted young Black men like me. Many young men from minority communities faced higher rates of conscription due to limited access to educational and occupational resources. In other words, individuals from minority communities often lacked the means to pursue deferments through full-time college attendance or secure certain types of employment exempt from the draft. Furthermore, some argued that the draft boards, responsible for selecting individuals for military service, were biased and more likely to choose candidates from minority communities.

The history of the *draft* remains one of the most controversial issues in the United States. Two-thirds of the US military who served in the Vietnam War volunteered for duty. The other one-third were drafted, primarily into the US Army. The Selective Service System during the Vietnam War was highly controversial, mainly because early in the war, draftees came disproportionately from poor, working-class, rural, and predominantly minority populations. Military conscription, commonly known as "the draft," is one of the most complex topics related to the Vietnam War. It remains an emotional sore point for many, including those who chose to serve when drafted, those who sought deferments to delay or avoid serving, and those who refused to serve and went to jail or left the country. Although there are conflicting statistics from different sources, it is clear that the draft was grossly unfair to minority segments of the population, especially early in the war. While it is officially acknowledged that the majority of those who served during the Vietnam War were volunteers, public backlash from draft-eligible men is a primary reason cited by many for why public sentiment later turned against the war effort.

The Vietnam War was not the first time in the nation's history that people were drafted. In fact, military conscription has been a regular practice in the United States since the American Revolution. When there weren't enough volunteers to meet the needs of the military, the Selective Service System, or the draft, was used to cover the shortfall. After World War II, the United

States maintained a "peacetime" draft, meaning the draft was already in place as America deepened its involvement in Vietnam. Local draft boards called up registered males and evaluated them for service. These boards were often made up of military veterans who granted exemptions and deferments for various reasons, such as medical issues and student status. Many perceived the deferment process as lopsided in favor of white males, especially those with financial privilege or political connections.

A 1969 revision of the Selective Service Act diminished the power of local draft boards and instituted a national lottery system. On national television on December 1 of that year, capsules containing birth dates were pulled from a large bowl. Those dates were then mounted on a large board in the order in which they were drawn. September 14th was the first date placed on the board, and in 1970, local draft boards began filling their quotas with draft-eligible men born on that date. Thousands of draft-age men refused military service in Vietnam. Burning draft cards, initially a symbolic protest, took on added significance in 1965 when President Johnson signed a law criminalizing the act. Some fled the country, often to Canada. A small number of men served jail sentences to protest the war. More than 3,000 men went to prison for some form of draft resistance. In all these instances, the overpoweringly resonant feeling was that Black men were severely discriminated against in all aspects of the draft.

After preliminary processing, I was inducted into the US Army. My first days as a *soldier* brought back recollections of my time in the high school Army Junior ROTC Program. Despite those memories, I continued to experience the gnawing feeling that the army was not my calling. As it happened, most of the soldiers in my platoon were to be assigned to units in Vietnam immediately after completing basic training. I had a premonition that most of my army comrades-in-arms would never return home from their tours of duty in Vietnam. The *Vietnam Veterans Memorial Wall* on the National Mall in Washington, DC, bears eloquent testimony to this premonition. Inscribed on the black granite walls are the names of more than 58,000 men and women who gave their lives for the war effort or remain missing. The Memorial honors the courage, sacrifice, and devotion to duty and country of all who answered the call to serve during one of the most divisive wars in US history.

I spent a week in the Army at Fort Holabird. Finally, unable to continue struggling against my innermost feelings of not being where I truly belonged, I approached the authorities with a request that I fully expected would be rejected. In a calm voice that starkly contrasted with my feelings of trepidation, I asked if I might be considered for a transfer from the Army to the Air Force. To my complete surprise, the authorities granted my request. Riding on a wave of unexpected euphoria, I left the Army and happily went to the Air Force Recruiting Station, where I wasted no time enlisting. That singular act of courage would significantly mitigate my earlier mistake of working full-time at my government job and irrevocably alter the trajectory of my life.

CHAPTER FIVE

INTO THE WILD BLUE YONDER

"If you want something that's going to provide you with a lot of challenges and a variety of different things to do, then you can't beat a place like the Air Force. I don't mean this to sound like a recruiting pitch. But it's been a lot of fun."
– Lieutenant Colonel Michael P. Anderson, NASA Astronaut

The *Wild Blue Yonder* is a distinctly American phrase. It alludes to a journey to a faraway location that is both appealing and mysterious. It refers to the open sky and describes a destination that is distant and indefinite. More significantly, for the benefit of the uninitiated, the phrase derives from the opening words of the official song of the U.S. Air Force, written by Robert Crawford in 1939. To clarify, the United States Air Force's official song is often called *"Wild Blue Yonder."* The first verse of the anthem encapsulates it all:

"Off we go into the wild blue yonder,

Climbing high into the sun;

Here they come zooming to meet our thunder,

At 'em now, Give 'em the gun! Give 'em the gun!

Down we dive, spouting our flame from under,

Off with one helluva roar!

We live in fame or go down in flame. Hey!

Nothing'll stop the U.S. Air Force!"

To further underscore the phrase's popularity within the United States Air Force, the official publication of Air University is *Wild Blue Yonder*. It is a peer-reviewed online journal and forum for military-related thought and dialogue. The journal aims to foster discussion and debate among practitioners and academicians.

So, having secured the unqualified permission of the Army authorities at Fort Holabird to transition from the Army to the Air Force, I flew off into the Wild Blue Yonder. To the ultimate glory of a fulfilling destiny that was about to open up to me, I officially enlisted in the US Air Force on August 10, 1965. This was not only a momentous occasion for me, but it would also mark a significant turning point in my life. I arrived at the Air Force Recruiting Station in Downtown Washington, DC, around noon that day to find about twenty-five prospective recruits in the office. Before being inducted into the US Air Force, one had to go through a process known as the *Military Entrance Processing Station (MEPS)*. The United States Military Entrance Processing Command (USMEPCOM) is a joint Service organization that determines an applicant's physical qualifications, aptitude, and moral standards as set by each branch of military service. They have MEPS locations all over the country.

At MEPS, I completed a series of tests and evaluations to determine my physical and mental fitness for military service. Although I had already completed all of the assessments during my short stint with the US Army at Fort Holabird, I still had to repeat all the required steps as a prerequisite for joining the US Air Force. This rather arduous process included a medical examination, completing the Armed Services Vocational Aptitude Battery (ASVAB), a criminal background check, and an interview with a career counselor. After completing all the tests and evaluations, I took the *Oath of Enlistment,* where I vowed to defend the United States Constitution and obey the Uniform Code of Military Justice (UCMJ). On that high note of patriotic fervor, I officially became an enlisted member of the US Air Force. Even after spending 35 years in the US Air Force, I still vividly remember reciting the solemn words.

"I swear that I will support and defend the Constitution of the United States against all enemies foreign and domestic; that I will bear true faith and allegiance to the same; and that I will obey the orders of the President of the United States and the orders of the officers appointed over me, according to regulations and the Uniform Code of Military Justice. So help me God."

After MEPS processing, I received my marching orders. I was given a date to report to Lackland Air Force Base in San Antonio, Texas, to begin training for my Air Force career. The Air Force had purchased airline tickets for my flight to San Antonio. I had some flight lessons in single-engine aircraft during my high school days, so my flight from Washington, DC, to San Antonio—my first time flying on a large commercial jet—was all the more fascinating. Needless to say, I found the flight exhilarating. I spent the entire time onboard the aircraft looking out the window and dreaming of someday becoming an Air Force aviator. After landing at the San Antonio Airport and claiming my baggage, I joined a group of twenty-five recruits waiting for transportation to Lackland Air Force Base. A non-commissioned Air Force officer warmly greeted us and ushered us onto waiting buses for the trip to Lackland. As soon as we arrived, a training instructor got on the bus and, with a sweeping look that allowed only the slightest hint of a regimental brand of aloof familiarity, began barking orders at everyone. At that moment, I realized that I was in the Air Force.

Every enlisted Airman begins their Air Force career with nine weeks of Basic Military Training (BMT). A mentally and physically challenged Airman acquires the skills and training to develop into a proper Airman. Air Force basic training is one of the most daunting challenges anyone can undertake. The Air Force prides itself on *excellence and perseverance* and places an uncommon, if not rare, premium on the extremely high standards it expects recruits to start meeting rapidly. Recruits are tested physically, mentally, and emotionally. Their morals are also tested, while their character is thoroughly scrutinized. Individuals are pushed harder and further than they have ever been pushed before. Naturally, this is because the Air Force is trying to find out if recruits have what it takes to join while teaching them the information, skills, and attitude needed to thrive in that branch of the United States Armed Forces.

As recruits, the adjustments we needed to make were vast. I soon discovered it was an entirely different way of life, driven by discipline, improvement, and a strong attitude toward every aspect of life and military service. As I observed, the lack of freedom was a challenge for many of us, as were the physical and mental difficulties of disciplined military life. Ultimately, I concluded that although Air Force basic training was challenging, it would eventually be one of my life's most formative and rewarding experiences. In any case, it granted me entry into a prosperous career with one of the finest military establishments in the world.

The hardest part of Air Force training probably differed for each of us in my recruit class, depending on the individual recruit. Some of us struggled with the regimentation that is so typical of military life. Initially, the lack of freedom and the emphasis on humility and discipline were hard to adjust to. I do not doubt that some of us found the rigorous physical training overwhelming, primarily since the demands of Basic Military Training were largely unanticipated. There was also the mental and emotional struggle of being away from one's family and friends and having to start all over again with a group of unfamiliar people with whom one was compelled to spend a lot of time.

Having established all these challenges, however, one week during the training was particularly dreaded. Coming somewhere around the middle of the training period, *BEAST* week was widely regarded as the most challenging phase of the training, and for good reason. *BEAST* is an acronym for *Basic Expeditionary Airman Skills Training*. It is essentially a mock *deployment,* putting into practice all of the training and skills one has learned up to that point. It was physically draining and demanding, requiring us to endure long hours of physical, combat, and tactical training. This training phase also included *Combat Arms Training and Maintenance (CATM)* and a fighting exercise that pitted recruits against each other. Yet, the highlight of BEAST week remained the Field Training Exercises, capping off the most rigorous essential training phase.

In summary, Basic Military Training at Lackland Air Force Base was both a challenging and rewarding experience that laid the foundation for my career in the Air Force. The nine-week training program consisted of four phases: *Reception, Foundation, Transformation,* and *Integration.* The initial week focused on uniform issues and medical screenings. Weeks 1-3 emphasized military discipline, customs and courtesies, Air Force Core Values, and basic military skills. Weeks 4-6 concentrated on physical fitness, close-order drill, and weapons familiarization. Weeks 5-9 included academic work, teamwork exercises, and deployment exercise scenarios, the last represented by *Basic Expeditionary Airman Skills Training.*

The sudden transformation to an unaccustomed sleep rhythm was one discipline the average enlistee had to adjust to quickly. In basic military training, there is no such thing as *sleeping in or* staying asleep in the morning for longer than usual. You had to get up at 5 a.m. every single day. Waking up in the morning was an adjustment process that was the same for every introductory training class. When we first arrived, the drill instructors required a lot of noise, yelling, and jostling to get everyone out of bed. Then, sometime around the middle of the training period, all it took was for the drill instructor to enter the room in the early morning and quietly say, *"Get up."* Everyone jumped out of their bunks immediately and began their morning routine. It was a genuinely remarkable adjustment. During the training, bedtime was usually 2100 hours, or 9 p.m., except during special events, such as night exercises. In basic training, lights out meant "*go to* sleep." It did not mean talking to your buddies, studying, or writing a letter home. Suffice it to say that these distractions generally did not pose a problem. In basic training, one always becomes so tired that falling asleep at night isn't an issue. What was more difficult was trying to keep from falling asleep during class time. As I recall, our chief drill instructor often spent the night with us in the barracks at the beginning of our basic training. However, as we progressed in training, the drill instructor would go home each night. Regardless, the essential training staff was constantly monitoring the barracks using closed-circuit cameras. If one was not quietly in one's bunk, one could expect a surprise nighttime visit from the drill instructor. All in all, we were never really without supervision.

I was a fast and diligent learner and had no problems completing basic training successfully. My previous military experience as a cadet in the high school Junior ROTC program served me well. It was a real asset that ultimately paid huge dividends during basic training. My training instructor, quickly recognizing my unusual discipline and potential for leadership, promptly designated me as a *Squad Leader.* I must point out, however, that I encountered several challenges in this role. As might be expected, my fellow Airmen came from a diverse array of geographical regions and included a wide range of backgrounds, including Black, White, and Hispanic individuals. Initially, due to cultural differences and the ingrained prejudices of some individuals in my unit, there were challenges with obeying orders and working as a cohesive team. However, as the weeks passed, a consensus arose within the group that the only way to complete basic military training was to work seamlessly together, *"Cooperate and Graduate,"* as it were. At the graduation ceremony, I was recognized as a *"Distinguished Graduate."*

An essential aspect of the Basic Military Training process was the interview with the Air Force Career Counselor. Based on test scores and an evaluation of my performance, the Air Force allowed me to select a career path or specialty that would align with both my desires and the needs of the military. I expressed interest in assignments in either the Electronics or Intelligence career fields during my interview. However, to my utter amazement, I was only offered jobs as a Security Policeman or a Combat Engineer. The counselor informed me that my first duty assignment would be in Vietnam if I accepted either of these jobs. I pleaded with the counselor that, based on my high aptitude scores and excellent performance evaluations, I should be considered for other options for my Air Force career. He finally relented and said I was qualified to take the language aptitude test due to my aptitude for foreign languages such as French, Latin, and Spanish. The test was in a dialect of Yiddish and involved learning the grammar, vocabulary, and syntax. According to the Air Force, *"If you could pass this test, they could teach you any spoken language."* During the war, there was a high demand for linguists who could speak Chinese, Russian, or Vietnamese. I passed the aptitude test with flying colors and was sent on a six-week Chinese language familiarization course. Unfortunately, I struggled significantly in my attempts to learn the Chinese language. Chinese is a tonal language, meaning that the tone or pitch of a word can change its meaning. To my dismay, it turned out that I was tone-deaf and could not correctly distinguish the tones when attempting to comprehend oral speech. After my failure with the Chinese language, the Air Force, in its infinite wisdom, decided to send me to the Defense Language Institute's East European Language School at Syracuse University for a nine-month intensive program in Russian, marking the beginning of my lifelong affiliation with the Russian language and culture.

CHAPTER SIX

LEARNING BASIC RUSSIAN AT SYRACUSE

"Russian is a very deceptive language because it looks easy at first. It's like setting out for a gentle stroll and realizing that you've committed yourself to scaling Himalayan peaks."
- Armand Hammer

Close to my utter failure at gaining linguistic familiarity with even rudimentary Chinese, the Air Force enrolled me at Syracuse University to attend the nine-month intensive introductory Russian Language Training Course in Upstate New York. In addition to Chinese, Russian is one of the most complex languages for English speakers due to its intricate grammar rules, various forms of verb conjugation, and, most significantly, its radically different alphabet, the *Cyrillic* script. The Cyrillic alphabet, a writing system developed sometime between the 9th and 10th centuries CE for Slavic-speaking peoples of the Eastern Orthodox faith, is currently used exclusively or as one of several alphabets for more than 50 languages across Eurasia. In fact, as of 2019, around 250 million people in Eurasia used Cyrillic as the official script for their national languages, with Russian accounting for half of them.

I arrived in Syracuse on November 1, 1965, and reported to the Commander of Detachment 1, 3345 Technical School, Syracuse University, to begin my language training. Syracuse University, a private research university in Syracuse, New York, was founded in 1870 when Genesee College, located in Lima, New York, and operated by the Methodist Church, relocated to Syracuse, where it began holding classes in 1871. Organized into thirteen schools and colleges, it has been a nonsectarian institution since 1920. The Air Force Detachment at Syracuse was part of the Defense Language Institute's East Coast Branch. This branch was established in 1963 as part of the Department of Defense's effort to provide intensive language training to military personnel. It focused explicitly on instruction in Russian, Arabic, Chinese, and Korean. The primary goal of the East Coast Branch was to equip Air Force personnel with the language skills needed for their assignments in intelligence, diplomacy, national security, and other military roles.

Attending Syracuse University gave me the unique opportunity to receive high-quality language training in a university setting. Furthermore, I was able to benefit from the academic resources and environment of the university while meeting the rigorous language training standards of the Defense Language Institute. I must admit that at the beginning of my Russian language training, I felt anxiety and apprehension and seriously doubted my ability to complete the course successfully. Fortunately, the deep and insightful dialogue I'd had with my mother when I was barely eight years old left a lasting impression on me, and I drew tremendous inspiration from her declaration that, with determination, perseverance, hard work, and a belief in God, I could accomplish virtually anything I aspired to in life.

Upon arrival at the language school, I found that all the rooms in the dormitory were already fully occupied.

As an alternative, I was assigned housing in a fully furnished, two-bedroom apartment in the New Slocum Heights Married Students housing complex, adjacent to the school and approximately a mile from the Syracuse main campus. For an airman who had just completed basic training, I was astounded and impressed by my luxurious accommodation. Only in the Air Force can one receive such glittering dividends at the incipient stage of one's career. In my basic Russian course class of fifty students, the only two African Americans were a fellow airman named Elmer and me. Elmer, who later became my roommate, had grown up in the inner city of East Baltimore. Naturally, that meant Elmer and I were coming from similar backgrounds, as we both grew up in the "hood," so to speak. We would regularly trade stories comparing our childhood experiences in the inner cities of Washington, DC, and Baltimore, Maryland. Quite a remarkable personality, Elmer was one of the most intelligent individuals I had ever known. He had been valedictorian of his high school class and had scored the top grade on the Air Force Language Aptitude Test. Getting to know Elmer and studying with him during our nine-month tenure in the Basic Russian Language course was a source of inspiration and joy for me, and we would become lifelong friends.

As I mentioned, in addition to Chinese, Arabic, Japanese, and Korean, Russian is one of the most difficult languages for English speakers to learn, mainly because it features the complex Cyrillic alphabet writing system and an equally complex verb conjugation system. Not surprisingly, my first daunting challenge in studying Russian was mastering the Cyrillic alphabet. Fortunately, I found striking similarities with Latin, which I had previously studied for six years in middle school and high school, making it relatively easy for me to master the Cyrillic alphabet. Russian is also considered a phonetic language because its writing system closely corresponds to the pronunciation of its words. Ultimately, my previous language studies in Spanish and French helped me quickly develop my conversational skills in Russian.

The curriculum of the basic Russian course was designed to provide future linguists with a foundational understanding of the language and prepare them for more advanced study. The course covered the following areas:

1. *Russian alphabet and pronunciation*
2. *Basic vocabulary and grammar concepts*
3. *Introduction to Russian culture and customs*
4. *Listening and reading comprehension exercises*
5. *Conversational practice*
6. *Writing and composition skills*
7. *Language drills and exercises*
8. *Russian language history and structure*

Formal classes at the Russian language school were held from 9 a.m. to 4 p.m., Monday through Friday. A maximum of ten students were in each class section to enhance the learning process.

Additionally, all students were required to complete at least two hours of homework each night. Regretfully, due to our heavy class and homework schedule, I was not afforded free time to meet and socialize with the regular students on the main Syracuse campus.

One of my primary outlets for extracurricular engagement and entertainment was basketball. I frequently went to basketball courts near the school for pickup games. One day, I met a fellow named Jim on the court. He was a Civil Engineering PhD student at Syracuse, and we lived in the same housing complex. Jim was an excellent basketball player, and we played together in many neighborhoods around Syracuse. Remarkably, we frequently won against much taller and more athletic opponents, most of whom were African American. Even more noteworthy about this narrative is that Jim was White and Jewish. He often referenced the movie *"White Men Can't Jump,"* intending to counter that assertion by telling me, *"I am White, and I can jump."* Jim completed his doctorate at Syracuse and later became a distinguished Professor of Civil Engineering at Syracuse University. I have known Jim for over fifty years and consider him one of my closest friends. I remain eternally grateful for the uncommon friendship, kindness, and hospitality that Jim and his family displayed towards me as a young Black airman.

I excelled in both written and oral examinations during the nine-month intensive Russian language program. For some inexplicable reason, learning Russian was an unexpectedly easy endeavor for me. On a slightly amusing note, I have often wondered whether I have some degree of Russian ancestry in my DNA. Just before graduation in August 1966, the school academic director invited me to his office and wholeheartedly congratulated me on my outstanding performance during the course. He also informed me that I had earned the distinction of being selected valedictorian of my class. This was an extraordinary honor, as I would be the first African American in the school's history to be named valedictorian.

I delivered my valedictory address at the graduation ceremonies held at Hendricks Chapel on the Syracuse University main campus to a packed audience of about three hundred. I delivered my remarks entirely in Russian, while English translations were passed out to the audience. This was one of the proudest moments of my life, and I was truly blessed that my mother and father had made the long trip from Washington, DC, to Syracuse, New York, to attend this significant milestone in their son's life. Shortly after completing my course at Syracuse, I received a call from the Air Force Assignments Officer. Due to my stellar performance in the Russian course at Syracuse, I entertained visions of obtaining a follow-on assignment in a warmer geographic location in the United States, such as Texas, California, or Florida. Unfortunately, the famous saying, *"If wishes were horses, beggars would ride,"* became my reality. To my dismay, after completing additional technical training, I was only offered an assignment at Fairbanks, Alaska, the coldest state in the United States. I set out for the continuation of my journey.

CHAPTER SEVEN

HOMECOMING

"I am the mother of an American Airman. I give my complete and unwavering support to my Airman. As my son serves the people of the United States, so I humbly offer up my prayers for his safety and the safety and health of those he serves beside. I accept that my Airman's first duty is to his country, and I understand that this sacrifice he willingly makes is what keeps our nation great. I know that an Airman's heart is true and strong and that my Airman will endure."
- An Airman's Mother's Creed in "My Daughter - An American Superhero." **By Marcy Pugh**

After completing the Basic Russian Course at Syracuse University, the next step in my Air Force journey was attending an eight-week technical training course in Texas to become an Airborne Communications Specialist. In this role, I would be responsible for operating and maintaining communications systems on aircraft. My duties would also include setting up and monitoring communications equipment, troubleshooting technical issues, and ensuring communication links were secure and functioning correctly during flight. Finally, I could fly on Air Force aircraft, not as a pilot or navigator, but as an aircrew member.

Now that I had satisfactorily fulfilled my immediate training obligations to the Air Force and fully established my status as a specialized Airman in the United States Air Force, it was time to take a well-deserved thirty-day leave. I returned to Washington, DC, to visit my parents and friends. All things considered, it was a tremendously joyous homecoming. Many friends were proud to see me in my new Air Force Service Dress Uniform. While I was home on that vacation, my mother insisted that I wear my uniform to church on Sundays so that she could proudly introduce her son, the Airman, to the entire congregation. I was more than happy to oblige my mother with her perfectly understandable desire to express pride in her son's achievement. Coming from a disadvantaged background as a young Black American, especially in an era when expectations for a fulfilling career in practically any field were relatively low for the average African American, my total identification with a My mother, who had always been a pillar of inspiration and support, came to me with infinite ease. Anytime I donned my Air Force Service Dress Uniform just to see her beam with pride, I was reminded of the profound dialogue I had with her when I was barely eight years old, during which she imparted a declaration that left a lasting impression on me: that I could accomplish virtually anything I aspired to with determination, perseverance, hard work, and a belief in God.

While many servicemen and women have been rightly honored for their sacrifices, I believe that a larger group of *unsung heroes* are the parents who also sacrifice by sending their children into harm's way. I want to express my gratitude to every parent who has experienced the anxiety of having a child in active military service, and especially to my mother, who faithfully and relentlessly nurtured my dreams until they came to fruition. None of us will become perfect in a day, a month, or a year. We might not even accomplish all we aspire to in our lifetime, but we can, at least, begin now, starting with our most obvious weaknesses and gradually converting them into strengths as we move forward on our journey. The quest may be a long one; it will be a lifelong endeavor, inevitably fraught with mistakes, with one falling and getting back up again, lifted by the power of ideals, self-worth, and determination to succeed. It will also demand tremendous effort. But we must refuse to sell ourselves short. We must make a little extra effort at each stage of our journey.

After spending time with my family, the next item on my itinerary was visiting my high school sweetheart, Vivian. Vivian was a beautiful, intelligent, refined, and fun-loving woman. Elegantly put, Vivian was the *"Girl of My Dreams."* After graduating high school, she worked as a fashion consultant at a high-end clothing store in Northwest Washington and also did some part-time modeling. Vivian lived with her parents in the upper-middle-class neighborhood of Beltsville, Maryland, a small town just outside Washington, DC. Located in Prince George's County, the city was home to many diverse cultures that united to create a vibrant and welcoming atmosphere. Gradually becoming a popular destination for shopping and dining, the small town where my beloved Vivian lived was ideal for those seeking an urban lifestyle with a rural feel.

Vivian's father was a media distributor, while her mother was an accountant. When I called Vivian to invite her to dinner to celebrate my homecoming, she readily accepted. We went to an expensive Italian restaurant in Georgetown. Shortly after dinner, during dessert, I got down on one knee, pulled out a diamond engagement ring from my pocket, and asked Vivian to be my wife. She looked me straight in the eyes and said, *"Yes."* We kissed passionately to seal our commitment to each other. Suffice it to say, it was one of the happiest days of my life. Because of the short duration of my stay in Washington, there wasn't sufficient time to plan a formal wedding before reporting to my next duty station. On December 27, 1966, we decided to elope and go to a Justice of the Peace in Washington, DC, to become husband and wife. As a matter of protocol and courtesy, we called our parents to inform them of our decision to marry. To our immense relief, they were all overjoyed. After the wedding ceremony, we booked a luxury hotel in Baltimore, Maryland, for a fabulous two-day honeymoon. Upon returning to Beltsville, Vivian's parents welcomed us warmly and said we could stay with them until I left for my next assignment in Fairbanks, Alaska.

CHAPTER EIGHT

OFF NORTH TO THE LAST FRONTIER

"Having now driven the majority of the fabled Alaska Highway, beginning at 'Mile 0' at Dawson Creek, BC, I've gained an appreciation for the massive feat of engineering and sheer human toughness it represents. I've always been interested in stories at the edge of the human experience, things that are just unbelievable by today's standards. Usually, these are tales of people taking on the kind of personal risk and sacrifice that we just don't understand in today's modern societies. On November 20, 1942, the Alcan Highway was officially opened."
- The Travel blog,edveturism.ca

Covering more than 1,500,000 sq km, Alaska is the largest U.S. state by area, the third least populous, and the most sparsely populated of the 50 states in the union. Its rugged landscape and climate, coupled with its low population density, have earned it the nickname '*The Last Frontier.*' The United States purchased Alaska from the Russian Empire on March 30, 1867, for 7.2 million U.S. dollars. The area underwent several administrative changes before becoming organized as a territory on May 11, 1912, and it was admitted as the 49th state of the U.S. on January 3, 1959. Population estimates in 2015 put Alaska's population at almost 740,000, with approximately half of those living within the Anchorage metropolitan area. Alaska's economy is dominated by the fishing, natural gas, and oil industries, which are in abundance. Military bases and tourism are also significant parts of the economy. The next step in my career as an Airman was a duty posting to one of these military bases, Eielson Air Force Base (AFB), located approximately 26 miles southeast of Fairbanks, Alaska. But first, I had to attend to a most compelling matter.

After all the happiness and euphoria of finally being married to Vivian, *"the Girl of My Dreams,"* the reality of my new circumstances took center stage in my consciousness. I could now step back to soberly and rationally consider the immediate and future implications of the significant leap into matrimony that we had just taken. The Army has an old saying: *"If we wanted you to be married, we would have issued you a wife."* In my opinion, the Air Force likely shares the same viewpoint. Yet, there was nothing to suggest that the armed services discouraged young military marriages. In fact, according to a Department of Defense study, nearly half of active-duty military personnel marry before they are 25, and two out of three are married before the age of 30. This is dramatically different from the general population, where the average age for a woman's first marriage is 27, and a man's first marriage is 29. Researchers also found a tendency for young couples to marry before one of them leaves on a military assignment. Various reasons have been offered for this trend.

Unlike civilian personnel, soldiers are not in training for years and don't take the approach that they should wait until their career is established before settling down. Many active duty personnel, especially those in combat zones, look to marriage to provide stability when the rest of their lives are unpredictable. The sense of loss that many soldiers experience after the death of fellow soldiers in combat gives them a feeling that life is tenuous and encourages them to be more proactive about the things they want to do with their lives, with marriages and families quickly becoming one of those priorities. Additionally, many soldiers want to be remembered and desire children as a legacy if they don't return home. Be that as it may, although Vivian and I were madly in love, I still felt a sense of apprehension about what the future held for us as newlyweds. I was faced with a multitude of troubling questions. As a lowly Airman in the early stages of a military career, my financial resources were rather precarious, to say the least. How on Earth was I going to support a growing family? Furthermore, Vivian had a promising career as a fashion consultant and model. Why would she give up on her dreams and aspirations for the future just to become my wife? Little, however, did I know just how fortunate a man I had become, as Vivian would eloquently prove.

One night, Vivian and I talked candidly about our future together. What she told me was staggering in its avowal of matrimonial fealty. Vivian expressed her deep love for me and that, as my wife, she was more than ready to accompany me to the ends of the Earth, Alaska included, should that be my next military destination. All that mattered, she added, was that we remained together. What was more, "The Girl of My Dreams" stated that her future career was not even an issue in the equation since her ultimate goal in life now was to be a good wife and, one day, a mother who would love, support, and respect her husband through both good times and bad. In the rather positive circumstances, we were left with no other mutual obligation than to sincerely profess our love for God and each other, and in the process, jointly reaffirm a portion of our marriage vows: "For better or worse, in sickness and health, till death do us part."If, up to this point, my marriage to Vivian had been the happiest event of my life, this very frank discussion was the most heartwarming dialogue I had ever experienced with anyone.

Now that my conversation with Vivian had yielded such an encouraging outcome and I felt I had established a solid foundation for our future together, I was in the right frame of mind to plan my trip to Fairbanks, Alaska. The Air Force had given me two options for travel to my new duty station: I could purchase an airline ticket to fly to Fairbanks from Washington, DC, or drive to my latest assignment. Being young, adventurous, and somewhat foolhardy, I exuberantly selected the second option. I drove alone across an expansive swath of the United States and a portion of Canada before traversing the Alaska Highway to my final destination, Fairbanks, Alaska. The Alaska Highway, also known as the Alaskan Highway or the Alcan Highway, is the road that connects Dawson Creek, British Columbia, in Canada, to Delta Junction, Alaska, passing through the rugged northern landscape of British Columbia, Yukon, and Alaska. The Alaska Highway was the engineering marvel of the World War II era and was once described as the largest and most challenging construction project since the Panama Canal. The highway, often referred to as the Alcan Highway (a shortening of "Alaska and Canada Highway"), was built in 1942 as a result of the bombing of Pearl Harbor the year before. The American government and military feared an overland invasion by the Japanese of the Aleutian Islands, which sit just 1,000 miles across the Pacific Ocean from Japan. As initially conceived, American engineers planned and constructed the Alaska-Canadian Military Highway to provide a central route from the contiguous United States into Alaska in the event that Japanese forces entered on foot.

The driving distance between Washington, DC, and Fairbanks, Alaska, is approximately 4,700 miles. Since it would be impossible to drive this considerable distance non-stop, I chose to break up the journey into multiple stops at various locations along the route to rest and take in the sights. I estimate it would take seven to ten days to reach my final destination. Before embarking on this adventure, I had to do some detailed planning. My first objective was to purchase a reliable automobile that would prepare me for any challenges that might arise during the long trip to Fairbanks. I was fortunate to find a 1960 Ford Galaxy at an auction. After a thorough mechanical inspection, I purchased the vehicle. Additionally, since Fairbanks, Alaska, was known to often record temperatures as low as minus seventy degrees Fahrenheit during the winter months, I had to shop for items that would protect me against such extreme weather. I stocked up on winter tires and snow chains, a block heater to keep the engine warm during cold temperatures, an engine oil heater, a battery blanket, and a detailed Winter Emergency Kitthat included essentials such as extra blankets, food, water, flashlights, and additional clothing.

After visiting my parents and friends and saying a tearful goodbye to my wife, Vivian, on February 1, 1967, I set off on my journey to Alaska. As I had to pass through many of the lower 48 states, I felt a sense of pride and awe at the stunning beauty of the land mass known geographically and geopolitically as the United States of America. I traversed various landscapes, from the rolling plains of the Midwest to the majestic mountains of the Rockies. For a young Black Airman who grew up in the slums of Washington, DC, and had never traveled outside of

Texas and New York, driving to Alaska was an adventure like no other, if not the adventure of a lifetime. Yet, a significant setback was finding suitable accommodation as I traveled across the United States. In that year, 1967, three years after the signing of the Civil Rights Act of 1964, Jim Crow laws were still in effect in several states to perpetuate racial segregation. These laws allowed businesses, including hotels, to deny access and services to Black people. On more than one occasion, I was forced to sleep overnight in my vehicle at rest stops along the highway due to a lack of suitable hotel accommodation. However, I did not let this obstacle hinder me from reaching my goal. In the back of my mind, I kept recalling my mother's words about persistence and determination.

Despite the extreme weather conditions, often plagued by heavy snowfall and limited visibility, my trip up the Alcan Highway was largely uneventful. In seven days, I arrived at Delta Junction, Alaska, approximately 98 miles southeast of Fairbanks. I drove the remainder of the distance the following day and arrived in Fairbanks. On the eighth day of my journey, I finally reached Eielson Air Force Base (AFB), approximately 26 miles southeast of Fairbanks, where I reported to my new duty station commander. After administrative processing, I was assigned to a room in the barracks on the base. At this point, the luxurious quarters I had occupied while at Syracuse University suddenly felt like a distant memory.

CHAPTER NINE

AN AIRMAN IN ALASKA

"I joined the Air Force. I took to it immediately when I arrived there. I did three years, eight months, and ten days in all, but it took me a year and a half to get disabused of my romantic notions about it."
- Morgan Freeman

I arrived in Alaska prepared to become a resident of what is famously known as the coldest state in America. With an average temperature of 26.6°F, which would often plummet to a low of -30°F during the winter months, my arrival at Fairbanks, Alaska, at the end of the first week of February 1967 couldn't have coincided with a colder phase of winter in America's famous *Last Frontier.* Nestled in North America's extreme northwest corner, Alaska's proximity to the North Pole is why winters here are long, and summers remain cold and cloudy. The Fairbanks area, in particular, experiences temperature extremes, with summer highs reaching 90°F and winter lows plunging to around -50°F. The severity of Alaska's cold climate is linked to ocean currents and the influence of winds from northern Canada and Siberia. In fact, the state's climate caused intense storms to sink boats anchored in Alaska's numerous harbors and triggered avalanches severe enough to close roads, schools, and government offices. Yet, amidst its icy trials, Alaska's beauty and cultural experiences offered me a great opportunity for personal growth and full integration into military life.

After settling into my *"luxurious"* single-room accommodation at the Enlisted Airmen's barracks on Eielson Air Force Base, I immediately called Vivian to inform her that I had arrived safely in Fairbanks. She told me that she had prayed daily for my safety during the long and treacherous 4,000-mile journey to my new duty station. After sharing the latest news from back home, my wife announced that she had something important to share. That was a very *pregnant* moment; no pun intended. I was filled with an anticipation I couldn't quite contain. I blustered with eagerness for her to disclose this essential piece of news. Vivian just laughed and told me to calm down. Then, she informed me that she had just returned home from a doctor's appointment. The good news was that she was pregnant with our first child. In other words, we would soon become parents.

I was ecstatic with joy. I was filled with indescribable excitement. I immediately told Vivian I was thrilled that we would start a family, and I promised her I would do everything I could to ensure we were reunited in Fairbanks before our baby was born. Throughout that day, I floated in sheer bliss on my cloud of elation. That I was ecstatically happy was undeniable. Having established a potentially fulfilling military career and successfully pulled off the secret of a clandestine marriage to *"the Girl of My Dreams,"* the next logical frontier for me to conquer was to start raising a family. When my conversation with Vivian ended, I sat in a chair for serious self-reflection.

I was a junior enlisted airman just starting my first Air Force aircrew member job. Additionally, at the ripe old age of twenty-two, I had a new wife and a baby on the way. I admitted that those were two significant undertakings at the dawn of a young man's life and career. However, by the grace of God, I knew that I had the combined inner strength and resilience to accomplish both tasks with creditable aplomb. The next day, I met with my commander and proudly told him I would soon be a father. He warmly congratulated me and promised to do everything he could to facilitate my pregnant wife's joining me in Fairbanks as quickly as possible. However, my commander also said that, according to our unit's worldwide reconnaissance mission in the ongoing Vietnam War, I would have to complete three Aircrew Survival Training Courses before joining a flight crew. The three courses were *Basic Aircrew Survival Training, Water Survival Training, and Arctic Survival Training.*

I first attended *Basic Aircrew Survival Training* at an Air Force Base in Spokane, Washington. The extensive training focused on teaching aircrew members the survival skills and techniques required if they found themselves in a downed aircraft or behind enemy lines. The course curriculum consisted of *shelter building, finding sources of food and water, land navigation, signaling for help, and evading capture if necessary.* Although the training was rigorous and physically and mentally challenging, it adequately prepared me to handle various survival situations in my later years as an aviator.

Next, I attended *Aircrew Water Survival Training* at an Air Force Base in Florida. This training is crucial for flight crews if they have to eject from or ditch an aircraft flying over a body of water. The training included practice sessions in pools and simulated open-water environments. The course curriculum covered techniques such as *ditching, escaping from a submerged aircraft, using life rafts and survival gear, signaling for rescue,* and *ways to conserve energy and stay afloat.* One of my most memorable experiences during water survival training was being extracted from the water by an Air Force search and rescue helicopter. I remember holding on to the hoist until I was safely inside the helicopter.

Lastly, I attended *Arctic Survival Training in* northern Alaska. The Air Force couldn't have chosen a better location for that training. With a minimum average temperature of about -17 degrees F, Alaska North is the coldest part of the United States. Given that our aircraft were stationed near Fairbanks, approximately ninety miles from the Arctic Circle, this survival course was essential for aircrew members to enhance their skills in survivability during a crash landing or related emergency in the Arctic wilderness. The training included *building a shelter in snow, starting a fire in icy conditions, finding and purifying water in snowy landscapes, navigation,* and *identifying and avoiding dangers such as frostbite and hypothermia.* Training in these harsh conditions prepared me for real-life scenarios later in life. It equipped me with the necessary skills to endure and overcome the challenges of the Arctic environment. Upon returning to Eielson AFB and completing my survival training, I was awarded my Basic Aircrew Member Wings and assigned to a reconnaissance flight crew to begin my career as an aviator. Ever since taking my first aviation course in high school, I had dreamt of being certified to fly on a military aircraft. To say it was an incredibly joyous occasion would be putting my elation rather too mildly.

My next task was locating suitable housing in Fairbanks so Vivian could join me at my new duty station. In 1967, Fairbanks had a population of around 20,000 people. It was a hub for Army and Air Force personnel stationed at Fort Wainwright and Eielson AFB. The town also hosted the campus of the University of

Alaska at Fairbanks. The city was also a center for mining and transportation, with the Alaska Pipeline undergoing construction at that time. Housing in Fairbanks was at a prohibitive premium and in short supply, primarily due to the increased numbers of military personnel and civilian workers flooding the city.

The housing situation in Fairbanks worsened on March 12, 1967, when a magnitude 6.4 earthquake struck Fairbanks. It caused significant damage to buildings and infrastructure in the area. Immediately after the quake subsided, I called Vivian to tell her I was not injured. However, I clarified that her arrival in Alaska would have to be unavoidably delayed by another two to three months until conditions in Fairbanks improved. Finally, in late June of 1967, I succeeded in renting a two-bedroom apartment in town. Vivian finally arrived at Fairbanks Airport on the Fourth of July, Independence Day of 1967. Words cannot express my utter joy in reuniting with my wife. Our reunion was also bittersweet because July 4th was Vivian's twenty-third birthday. After retrieving Vivian's luggage, we drove back to Fairbanks to our new residence and began the next chapter of our incredible journey.

CHAPTER TEN

FAIRBANKS - "LAND OF THE MIDNIGHT SUN"

"The sun rises at midnight."
- Sonia Delaunay

When I opened the door to our new apartment in downtown Fairbanks and escorted Vivian inside, I noticed my wife's priceless expression of surprise and amazement. I couldn't have been more delighted by the wholesome effect my efforts at decorating our new home had on her. I had placed two large banners on the dining room walls to ensure that her arrival was memorable. One announced, *"Welcome Home,"* while the other greeted her, *"Happy Birthday To My Wife."* I also set a vase with a bouquet of red roses on the dining room table next to a small birthday cake. Upon seeing all the effort I had put into making her homecoming special, Vivian dissolved into tears of joy.

After concluding our birthday celebration, it was time to retire to bed. Vivian was exhausted from the nine-hour flight from Washington, DC, to Fairbanks. To make matters worse, she had to endure this long journey alone while pregnant with our first child. It was well after 10:00 pm before we finally went to bed, and although midnight was fast approaching, it was still very bright outside, with the sun's rays streaming into our bedroom.

"Why is it still so bright outside, like noon?" Vivian asked in surprise.

"Honey, the regions near the Arctic Circle, including Fairbanks, are commonly referred to as 'The Land of the Midnight Sun' because the sun remains visible for almost twenty-four hours a day during the summer months," I responded.

I also explained to Vivian that going to sleep could be quite challenging because sunlight constantly streamed into one's living space. As a remedy, I placed aluminum foil over all the windows to block the light.

Fairbanks has long been known as *"The Land of the Midnight Sun."* Alaska is home to many seasonal phenomena, and while the Alaskan winter is also known for the *northern lights,* no other seasonal phenomenon quite rivals Alaska's summer midnight sun. Endless summer days give meaning to the term *midnight sun,* which refers to up to 24 hours of sunlight; much of Alaska enjoys this from the end of April through mid-August, with the longest day of the year being the *summer solstice on* June 21. During the Midnight Sun Season, the sun never seems to set. You may wake up in the middle of the night to find the sun shining brightly and people out biking, gardening, and walking their dogs. Remarkably, residents of Fairbanks celebrate the sun with various events during this period. Indeed, the endless sunshine transforms life in Fairbanks in pleasing and wondrous ways. Golfing in the early hours of the morning is commonplace. Flowers, vegetables, and all things *green* and *growing* practically dwarf those in other American states in size. Additionally, the sun energizes people to be active at all hours, and their favorite late-night activities include hiking, running, golfing, and even baseball.

The summer days in Fairbanks are so long that even locals lose track of time, with many hotels resorting to blackout curtains to block the midnight sun. The midnight sun season holds real significance for Alaskans, offering endless hours to enjoy the outdoors and make up for the short winter days, along with producing award-winning giant garden produce. Alaskans also like to celebrate the midnight sun and the joy, energy, and activity the season brings, with festivals and events throughout the state for locals and visitors to enjoy. In fact, both the winter and summer solstices hold sacred meaning for Alaskan natives. While the winter solstice symbolizes *new beginnings as* the days begin to extend and brighten, the midnight sun marks the start of the year's closing, as daylight hours become shorter and days grow darker after the summer solstice. There is, in fact, a deep-rooted connection to *ancestry, tradition, folklore, and religion* in these solstices that is celebrated across Alaska.

Despite all these seemingly bright dividends of the midnight sun, there are significant disadvantages to health and well-being. Further research on my part revealed that due to the extended periods of daylight during the summer months in Fairbanks, many residents experienced psychological problems, including *sleep disturbances, depression, mood changes, anxiety, restlessness, and social isolation.* Naturally, my immediate preoccupation became ensuring that my pregnant wife adjusted well to her new and unique environment. Indeed, all thoughts of the novelty of the midnight sun faded into the background as my primary concern became the health of my wife and our unborn child. This required constant trips to the Obstetrics Clinic at Fort Wainwright Army Hospital in Fairbanks for prenatal care. We also frequently drove the twenty-six miles from Fairbanks to Eielson AFB to shop at the base exchange and commissary, see a movie, socialize with fellow airmen and their families at the Base Non-Commissioned Officer's Club, and attend church services.

As fate would have it, one Sunday after church, we were introduced to two individuals who would profoundly affect our lives in Fairbanks. Everett was a Senior Non-Commissioned Officer who worked in aircraft maintenance, and his wife Joyce was an elementary teacher on the base. Everett and Joyce had been stationed at Eielson for four years and took Vivian and me under their wings. They understood the challenges of living in the Far North because they had experienced them personally. As a result of meeting this wonderful and caring couple, Vivian and I were able to expand our circle of friends and establish a support network that made our tour in Alaska much more enjoyable and manageable.

With a total population of about 20,000 residents, Fairbanks had the look and feel of a small Midwestern town. It did not offer much amusement other than visiting the one-level shops, restaurants, bars, and saloons in town. The J.C. Penney department store was the only multistory building in the city with an escalator. To my amusement, many residents of Fairbanks and the surrounding areas would come to town to shop and ride up and down the escalator on weekends. During the summer months, I would take Vivian to explore some of the popular sightseeing attractions in Fairbanks each weekend, including *Pioneer Park,* which showcased the history of Fairbanks and Alaska and featured museums, historic buildings, and displays; *Morris Thompson Cultural and Visitor Center*, which offered information about local attractions, events, and culture; *Riverboat Discovery Cruise* along the Chena

River, which provided views of the beautiful wilderness scenery around Fairbanks; and the *University of Alaska Fairbanks,* where we explored the campus and attended events, art exhibits, and lectures. The traditional Midnight Sun Baseball game was a unique attraction during the summer months. The game was played at night when the sun remained visible for nearly 24 hours. Although Vivian arrived in Fairbanks in early July after the traditional Midnight Sun baseball game had been played in June, we had the opportunity to attend a baseball game on the base in mid-July, where we watched two local teams play baseball in the middle of the night with the sun never setting. To call it a surreal experience would be an understatement.

Living in Fairbanks during the summer was pleasant due to the relatively mild temperatures and extended daylight hours. However, the winter months presented an entirely different picture because of the frigid temperatures, limited daylight, and heavy snowfall, making life in this arctic environment uniquely challenging. Most residents of Fairbanks agreed that *"if you could survive the harsh arctic winters, you could live anywhere, even on the Moon."* The average winter temperature in Fairbanks typically ranged from -20°F to -40°F, with temperatures sometimes dropping as low as -65°F. When venturing outside, we always had to be prepared with the appropriate winter clothing, such as insulated jackets, parkas, boots, and gloves. Additionally, ensuring that your car was in good mechanical condition and equipped with an engine block heater was essential.

Another factor we had to deal with during the cold, dark winter months was a psychological condition called *Seasonal Affective Disorder (SAD).* Symptoms of SAD include low mood, lack of energy, increased appetite and weight gain, difficulty concentrating, a desire to sleep more than usual, and feelings of isolation and loneliness. Since Vivian was pregnant during the winter months, I had to do everything possible to help my wife combat the effects of SAD.

Despite the harsh winter conditions, I still had to fulfill my primary duties as an aircrew member on a reconnaissance aircraft. I recall many days during the winter when my flight crew had to take off and land in total darkness, with temperatures reaching as low as -65°F. Upon landing and after completing ground maintenance checks and mission debriefings, I would drive my old 1960 Ford the 26 miles from Eielson AFB to our home in Fairbanks.

I vividly remember the events of the night of December 13, 1967. When I returned home from a long flight around 11:00 pm, Vivian greeted me at the door with a small overnight bag. She told me calmly that it was time to go to the hospital to have our baby. I tried to suppress the feeling of panic that threatened to overwhelm me. The temperature that night was 65°F, and it was snowing. After getting into our old car, I prayed to God for a safe trip to the Fort Wainwright Hospital Emergency Room. Luckily, despite the awful weather, we made it to the hospital in record time. Naturally, I was on pins and needles for about four hours until the doctor finally came into the waiting room and announced that I was the father of a healthy eight-pound baby boy. I rushed into Vivian's room and tenderly kissed her. Words cannot express my joy at finally having our first son. We named him Steven Alan Wallace. Thus, we began the next chapter of our incredible journey.

CHAPTER ELEVEN

SEASON OF MIXED BLESSINGS

"Things turn out best for the people who make the best of the way things turn out."
- John Wooden

After the birth of our first son, Steven, Vivian quickly adapted to her new role as a mother.

Seeing how wonderfully she bonded with our son, spending quality time with him brought me tremendous joy. However, my elation was tempered by certain concerns about my wife's well-being. My apprehension stemmed from the long periods I was away from home due to my hectic flying schedule. Much to my discomfiture, there were numerous times when Vivian was left alone in our tiny apartment in Fairbanks without a reliable support system to help her with Steven. I had little doubt that adjusting to motherhood would be an overwhelming experience for her, and I knew I had to take special care to ensure she did not experience long periods of stress, anxiety, or depression. As luck would have it, I was fortunate to find an excellent caregiver to assist my wife with baby Steven during our remaining time in Alaska.

After completing approximately fourteen months of my tour of duty at Eielson AFB, it was time to contact the Air Force Assignments Officer to learn about the decision regarding my next duty station. Having spent my first duty tour in Alaska, it was only natural that I would yearn for warmer climes, and I lobbied heavily for an assignment in one of the *"Lower 48 States,"* such as Florida, Texas, or California. To my utter surprise, however, the Assignments Officer informed me that due to my outstanding performance as a Communications Specialist and Combat Air Crew member, the Air Force had decided on my early promotion to the rank of Sergeant and to send me back to Syracuse University for the nine-month Intermediate Russian Language Course. This course, understandably, was not just fantastic news but also melodic music to my ears. The reason was clear.

When I told Vivian that I was being promoted early to the rank of Sergeant and being posted back to Syracuse University, she was, not surprisingly, elated. Syracuse was only a comparatively tolerable six-hour drive from Washington, DC, given our current location at America's last frontier. Additionally, Vivian was almost deliriously happy that she would be closer to our family and friends. To be selected for this prestigious assignment after serving less than three years in the Air Force was an exceptional feat. Yet, despite the sincere acknowledgment of the honor and privilege the Air Force was granting me, I still entertained mixed emotions about returning to school.

Just before we left Fairbanks, Vivian informed me that she was pregnant with our second child. I was, of course, elated that our second child was on the way. Our son, Marten David Wallace, was born on June 23, 1969, during my second tour at Syracuse University. Upon returning to Syracuse, my primary concern was how Vivian, Steven, and our newborn baby would adjust to being left alone for extended periods while I was in class, studying at the library, or completing my daily homework assignments. Once back in Syracuse, my first order of business was to locate a suitable apartment for our growing family near the university campus. Luck was obviously on my side, as I finally identified and rented a spacious three-bedroom apartment in a decent neighborhood about four miles from the main campus.

After settling the family, I reported to the Commander of the Defense Language Institute at Syracuse University to begin my *Intermediate Russian Language Course.* The Defense Language Institute at Syracuse University is essentially an outpost of the *Defense Language Institute (DLI),* a United States Department of Defense educational and research institution consisting of two separate entities that provide linguistic and cultural instruction to the Department of Defense, other federal agencies, and numerous agencies around the world. The Defense Language Institute is responsible for the Defense Language Program, and the bulk of its activities involve educating Department of Defense officers and officials in assigned languages and international personnel in English. Other functions include planning, curriculum development, and research in second-language acquisition.

The intermediate Russian course at Syracuse University continued language skills development in reading, writing, speaking, and listening comprehension. The curriculum also focused on more complex grammar structures, vocabulary expansion, and cultural understanding. All students were required to engage in discussions, presentations, and written assignments to further hone their proficiency in the Russian language. Because I had completed the Basic Russian Course at the top of my class and was designated valedictorian, I tried valiantly to repeat that feat in the intermediate course. However, due to competing priorities and the compelling obligations of raising a family, I finished the course as the second-highest performer and was designated a *Distinguished Graduate.*

A valuable lesson from my second tour at Syracuse University was that one cannot always be the top achiever. It is more important to be a well-rounded person and to put your family first, above all else. I also learned that, in your quest to be a high achiever, you are only ever as good as your last success. This can make you proud as well as insecure at the same time. There's always another mountain to climb, another round of applause to receive, and another platform to deliver the valedictory speech. It never ends. Ultimately, however, our most significant achievements lie in the deep human connections we can forge through the fundamental values that define our character and our purpose for living and achieving in the first place. As I humbly accepted my second place at the top of the intermediate Russian course, I was grateful for the even more remarkable accomplishments of my wife and two sons.

Halfway through my intermediate Russian language training, I received a call from my friendly Air Force Assignments Counselor. As with previous conversations, I again requested a follow-on assignment in a warmer climate, such as Florida, Texas, or California. The man was always full of surprises. To my amazement, he informed me that my next assignment would be a twelve-month remote tour, without my family, to Turkey. I told him I strongly opposed leaving my young family in the United States while I spent twelve months in Turkey. He relented and asked if I would consider returning to Alaska for a three-year tour. After a few moments of reflection, I decided that it would be wiser to deal with a familiar situation, even if it was imperfect, than to take a risk with an entirely new and unfamiliar set of circumstances that might turn out much worse. I immediately accepted the offer to return to Eielson AFB to resume my flying duties.

I had another, yet eminently attractive, incentive for returning to Alaska. Upon completing my language training, I was scheduled for promotion to the Non-Commissioned Officer (NCO) rank of Staff Sergeant. That rank would instantly qualify me for on-base housing at Eielson AFB. I couldn't think of a more favorable domestic arrangement for my young family. My euphoria about returning to my flying duties in Alaska would, however, suddenly turn to sorrow and grief when I learned that one of our reconnaissance aircraft at Eielson AFB had crashed over the Bering Sea on June 5, 1969. Despite an exhaustive search of the area, no wreckage and no survivors were found. The best theory was that the plane had suffered some catastrophic mechanical failure, as the last radio transmission from the aircraft mentioned *"vibration in flight."* To this day, it still haunts me that I lost twelve of my fellow aircrew members on that flight. Nevertheless, duty called, and I commenced preparations to return to Alaska to resume my flight duties in defense of our nation. My saga continues.

CHAPTER TWELVE

BACK TO THE LAND OF THE MIDNIGHT SUN

"One of the saddest things I've witnessed during my 35 years with the Air Force, including over 23 years active duty, as a contractor, and now as an Air Force civilian, has been people who stay with our service as a career, yet never use the many educational benefits available to them. My educational journey is, in many ways, a story about all the educational benefits that I've been fortunate enough to take advantage of over the years."
- **Mark Bacon,** Air Force Veteran

After completing the Intermediate Russian Language Course at Syracuse University, it was time to continue the next chapter of my incredible journey. As mentioned, the Air Force reassigned me to Eielson AFB, Alaska, to resume my aircrew flying duties. Initially, Vivian was slightly apprehensive about returning to the *"Land of the Midnight Sun."* However, I finally convinced her that, given the circumstances, this was our best option if we wanted to stay together as a family to jointly raise and care for our two young sons.

We left Syracuse and drove to Washington, DC, for a two-week leave to visit family members and friends before flying back to Fairbanks. Because I had received an early promotion to the Non-Commissioned Officer (NCO) rank of Staff Sergeant in the Air Force before leaving Syracuse, I was allocated a three-bedroom, two-bath townhouse when we arrived at Eielson AFB. Being accommodated on base made our lives so much easier, as we now had convenient access to all the base amenities, including the Commissary, Base Exchange, Movie Theater, Day Care Center, Gymnasium, NCO Club, and Church Services. I no longer had to drive the 26 miles from Fairbanks to Eielson daily for work.

As I mentioned, living on the base made our everyday experiences much easier. The Base Commissary is the neighborhood grocery store, and such a commissary is located on military installations all over the United States. The commissary sells food and household items at prices often below those of other grocery stores. On average, one can cut nearly a third off one's grocery bill compared to in-town prices. Additionally, one can save up to 50% or more during commissary special case lot sales. The Base Exchange, on the other hand, is a retail store, typically set up like a department store or a strip mall, with smaller shops and service vendors nearby. Many military installations in America have an exchange, some with uniform shops, barbershops, laundry and dry cleaning, gas stations, convenience stores, fast food outlets, and lawn and garden shops. Every service branch has its exchange system, like a corporate entity, similar to what is seen in the commercial world, such as Walmart, Target, and Macy's.

After completing my aircrew standardization examinations, I was again certified to fly as a Combat Aircrew Member. It was a tremendous joy to join a flight crew again. However, the downside was that I had to spend considerable time flying worldwide on temporary duty (TDY) assignments. Unfortunately, this left Vivian alone with our two little boys on base for extended periods. I wasn't too happy about this, although Vivian never, not once, complained about my protracted absences. As an Air Force wife, she recognized, much to her credit and even more to my sincere appreciation, her responsibility to support me in my mission to defend our country from foreign threats.

On one of my reconnaissance missions, while flying over the Atlantic Ocean, a group of birds smashed the cockpit window of our aircraft and caused a cabin depressurization. All flight crew quickly donned their oxygen masks to ensure their safety and the ability to carry out emergency procedures. However, the cabin depressurization resulted in zero gravity in the aircraft as the pilot initiated an emergency descent from 30,000 feet to 10,000 feet. Thanks to the pilot's skilled airmanship, we attained a safe altitude and controlled the situation. However, had the pilot been unable to stabilize the aircraft during the descent, we would have crash-landed in the ocean, killing everyone onboard. During the rapid descent, most of the flight crew were floating in zero gravity in the crew cabin. Unfortunately, one unlucky airman was in the aircraft toilet during this incident. The liquid waste from the chemical toilet engulfed his flight suit. When we finally got the aircraft back under control and landed at Eielson, this individual loudly exclaimed that this was his "last flight" as he solemnly took off his aircrew wings.

At the end of the first year of my second tour at Eielson AFB, my flight crew had the distinction of winning the *Combat Crew of the* Quarter Award. This great honor recognized our professionalism during reconnaissance missions. Additionally, each crewmember was decorated with the prestigious *"Air Medal,"* awarded to United States Air Force (USAF) personnel for meritorious achievements while participating in aerial flight. After completing two years of my three-year tour at Eielson AFB, I decided it was time for me to continue my college education, so I applied for the Air Force Operation Bootstrap Program. Operation Bootstrap was a commissioning program established in the 1960s. It offered a pathway for enlisted Air Force personnel to earn a commission to become officers. The program was designed to provide highly motivated and qualified enlisted airmen with the opportunity to become officers through an excellent combination of education, training, and experience. If selected for this prestigious program, I would attend college full-time. The Air Force would cover the cost of all tuition and fees and provide me with a monthly stipend to support my education. Upon graduation and completion of all program requirements, I would be commissioned as a United States Air Force second lieutenant.

As soon as I applied for the commissioning program, I received negative feedback from my fellow NCOs. In their view, it was not in my best interest to become an "officer" but to continue my Air Force career as an enlisted man. I rejected this argument and countered that the program would be an excellent opportunity for me to serve the Air Force at higher levels of command and responsibility. With the wisdom of hindsight, I can now understand the efforts of my peers to discourage me from seeking a commission. It is human nature that when one attempts to leave others behind on the ladder of life, one's peers are more than likely to, consciously or subconsciously, present all arguments, logical or otherwise, to discourage one from leaving them behind.

My application for the Bootstrap Commissioning Program was mysteriously misplaced twice for some inexplicable reason. However, whenever adversity and negative trends surface in any aspect of my life and career, something as inevitable as the fact that daytime will eventually give way to nightfall, I would always return to the soothing words my mother told me as a young child, *"Son, in this great country, you can be anything you want to be. It just takes believing in yourself, getting a good education, working hard, and never giving up on your dreams."* Finally, my application for the commissioning program was submitted and approved by the Air Force. I received orders to return to Syracuse University to complete my undergraduate degree in Russian Area Studies. My saga continued as I returned to Syracuse for the third time in my six-year Air Force career.

CHAPTER THIRTEEN

RETURN TO SYRACUSE UNIVERSITY

"Love of Learning is the Guide to Life."
- Motto of Phi Beta Kappa Society

After saying goodbye to all our friends and colleagues at Eielson AFB, on the first day of June 1971, Vivian, the two boys, and I departed Fairbanks, Alaska, for Syracuse, New York. In a career that was in its sixth year of smooth and fulfilling progression, I was returning to Syracuse University for the third time, this time to embark on the Bootstrap Commissioning Program that would position me for entry into the officer cadre of the United States Air Force. Previously, I had spent approximately two years at the Defense Language School's East Coast Branch at Syracuse, completing both the Basic and Intermediate Language Courses. This time, I had been accepted by the Syracuse University College of Liberal Arts as a full-time student to complete a one-year undergraduate degree program in Russian Area Studies.

Syracuse University is a well-established private research university with a strong reputation for academic programs and initiatives in various fields, including business, liberal arts, engineering, and the sciences. Syracuse is so focused on research that it is classified among *"R1: Doctoral Universities - Very High Research Activity."* Organized into thirteen schools and colleges, Syracuse University was established in 1870 with roots in the Methodist Episcopal Church but has been nonsectarian since 1920. Located in the city's University Hill neighborhood, east and southeast of Downtown Syracuse, the large campus features an eclectic mix of architecture, ranging from nineteenth-century Romanesque Revival to contemporary styles.

When I arrived in Syracuse in late spring to early summer of 1971, the campus was in the throes of activism and social change, as this was the era of social upheavals and protest movements sweeping through a large swath of the United States. At that time, the topics under fervent discussion by students and faculty included the Vietnam War, civil rights, and feminism. Having spent six years of military regimentation in the Air Force, mostly in Alaska, I had been somewhat insulated from the political and social turbulence that held America in its fevered grip in 1971. As an Air Force enlisted man and crewmember, my primary focus was the protection and defense of the United States against enemy aggression, both foreign and domestic. The words in the *Oath of Enlistment are clear and unambiguous about that mission: "I swear that I will support and defend the Constitution of the United States against all enemies foreign and domestic; that I will bear true faith and allegiance to the same; and that I will obey the orders of the President of the United States and the orders of the officers appointed over me, according to regulations and the Uniform Code of Military Justice. So help me God."*

Yet, being a Black undergraduate student at Syracuse University in 1971 presented a unique mix of challenges and opportunities. At Syracuse, Black undergraduate students across all academic disciplines accounted for about six percent of the total student population. During my final year at Syracuse, many Black students faced instances of racism and discrimination on campus. However, as I discovered—much to my delight—Black students at the university found a strong sense of community and solidarity within their small numbers through enthusiastic participation in Black fraternities, sororities, and organizations that celebrated Black culture and history. Additionally, many Black students were instrumental in advocating for more diversity and inclusion at the university. I vividly recall that in February 1972, a group of students organized a sit-in to protest racial discrimination and inequality on campus. The students occupied the administration building to demand increased representation of minorities in faculty and curriculum, as well as improved support services for minority students.

In my Russian Area Studies Program, I was the only Black undergraduate in a class of fifty students. However, due to my previous outstanding academic performance in both the Basic and Intermediate Russian Language Courses and my specialized training in Russian Language and Culture, I found myself excelling in all my coursework, culminating in a final grade point average (GPA) of 3.85. At graduation, I received the designation *Summa cum laude,* which translates to *"with highest honors,"* for my excellent academic performance. *Summa cum laude* is a Latin honor, and I consider it the high point of my academic career. Latin honors are a system of Latin phrases used in colleges and universities to indicate the level of distinction with which an academic degree has been earned. Primarily used in the United States, this honors distinction should not be confused with the honors degrees offered in some countries or with *honoris causa* degrees. The system has three levels of honor, listed in order of increasing merit: *cum laude, magna cum laude,* and *summa cum laude*. Generally, a college or university's regulations specify the criteria a student must meet to obtain a given honor. Being awarded my bachelor's degree with the honor of *summa cum laude* was the culmination of ten years of academic study. I graduated from high school in 1962 and obtained my undergraduate degree in 1972. For me, it had been a ten-year academic journey notable for its vindication of my combined resolve, persistence, and hard work. After graduating from Syracuse, I received orders to attend the Air Force Officer Training School at Lackland AFB, San Antonio, Texas. As a Distinguished Graduate, I completed the twelve-week Officer Pre-Commissioning Course and received my Gold Bars as an Air Force Second Lieutenant.

Despite my stellar academic performance at Syracuse, I was utterly disbelieving when the university did not invite me to join the *Phi Beta Kappa* academic honor society. Admittedly, membership in this organization is highly selective and considered a prestigious honor for undergraduates. Still, I was appalled by my non-invitation when some of my White peers, with lesser GPAs, were invited to join the society. Membership in *Phi Beta Kappa* is typically limited to students with very high GPAs, at least 3.8 on a 4.0 scale. I had scored a GPA of 3.85. The failure of the *Phi Beta Kappa* society to invite me into its fold raised a couple of obvious questions: *Was this a mere oversight? Was this blatant discrimination?*

Four years after graduating from Syracuse University, I wrote a letter to the president of Syracuse University is requesting an explanation for why I was not invited to join the Kappa Chapter of Phi Beta Kappa upon graduation. In his response, the president stated that I should have been asked to join this honorary society and would receive a formal invitation to return to Syracuse for the induction ceremony. Upon my eventual return to Syracuse for the induction ceremony, I felt a sense of vindication at the correction of the oversight, if not even restitution for the barely concealed contempt in which I had been, wittingly or unwittingly, held. Indeed, I felt a sense of joy and jubilation that, through persistence and determination, I eventually received the recognition that I rightfully deserved.

The *Phi Beta Kappa* Society is the oldest academic honor society in the United States. It was founded at the College of William and Mary in Virginia in December 1776. *Phi Beta Kappa* aims to promote excellence in the liberal arts and sciences and to induct outstanding students at select American colleges and universities. As the first collegiate organization to adopt a Greek-letter name, *Phi Beta Kappa* is generally considered a forerunner of modern college fraternities and the model for later collegiate honorary societies. The symbol of the *Phi Beta Kappa Society* is a golden key engraved on the obverse with the image of a pointing finger, three stars, and the Greek letters from which the society takes its name. On the reverse are the initials "SP" in script. The "S" and "P" stand for *Societas Philosophiae.* Since its inception, its inducted members include seventeen United States presidents, forty-two United States Supreme Court justices, and 136 Nobel Prize laureates. I insisted on taking my rightful place on this honor roll, and I took it.

CHAPTER FOURTEEN

BECOMING AN AIR FORCE INTELLIGENCE OFFICER

"No man succeeds without a good woman behind him. Wife or mother, if it is both, he is twice blessed indeed."
- Godfrey Winn

Having become a newly minted Air Force Second Lieutenant, my first order of business was to contact the Officer Assignments Branch to discuss the possibilities for my first tour of duty. As in the past, I had submitted the proverbial "dream sheet" listing my three assignment choices. An officer fills out a form, known as a "dream sheet," to indicate their assignment preferences. It is called a *dream sheet for* a very pertinent reason. Although the Air Force will consider preferences, the overriding factor is where the Air Force needs personnel the most. If the Air Force's needs fortuitously coincide with one's choices, one might consider oneself highly fortunate, and both the Air Force and the officer will be happy. If the Air Force's needs do not coincide with one's preferences, the officer will be assigned to where the Air Force needs them.

After serving six years in the cold climes of Fairbanks, Alaska, and Syracuse, New York, I made a renewed attempt to advocate for an assignment as an Operations Officer with an Airborne Reconnaissance Unit in Texas or Florida. However, my preferences were rejected by The Assignments Officer, who informed me that I was going to the 544th Air Intelligence Wing, based at Offutt Air Force Base in Nebraska, as an Intelligence Officer. The 544th Air Intelligence Wing is subordinate to the Strategic Air Command (SAC) and is responsible for supporting various missions, including providing intelligence to military operations, conducting analysis of potential threats, and assisting in mission planning. Additionally, the wing collaborates closely with other units within the Air Force, as well as with different branches of the military and various intelligence agencies, to gather and analyze information to support national security objectives. It plays a crucial role in ensuring the success of military operations by providing timely, accurate, and relevant intelligence to decision-makers.

She was not overly enthusiastic when I told Vivian about my new assignment. Although in the past, she had always claimed to be my "number one cheerleader" and would follow me "to the ends of the earth," this time, her disposition seemed remarkably different. I could detect the undertones of real disappointment in her voice. After two Air Force assignments in Alaska and New York, Vivian expressed that she was exhausted from moving from one cold climate to the next. Omaha, Nebraska, was located in the Midwest of the United States, and she didn't seem thrilled about relocating to an area known for its extreme contrasts of hot, humid summers and cold, snowy winters.

Vivian also stated that although she loved me very much and was committed to being a supportive wife and a good mother to our two boys, she now felt the urge to focus more on her own goals and pursue a career in the nursing field. Naturally, I responded that I would fully support her decision to further her education. In retrospect, I believe that I had been so fixated on advancing my Air Force career that I had failed to pay adequate attention to the needs and desires of my beautiful wife. Following our candid conversation about our plans, I immediately contacted the Air Force Officer Assignments Team to request a change in my assignment from Nebraska to another location. Regretfully, I was told in no uncertain terms that the immediate needs of the United States Air Force unequivocally required my assignment to Offutt AFB. Yet, the assignments officer proposed a compromise.

I could first go on a permanent change of station (PCS) assignment to Lowry AFB in Denver, Colorado, to attend the twenty-week Basic Officer Air Intelligence Course before beginning my tour in Nebraska. This would delay my assignment to Offutt, allowing me to move my family to Denver while I attended school. I quickly accepted this offer. In my view, proceeding first to a large, vibrant, and cosmopolitan city like Denver would be something that Vivian would approve of. Vivian was overjoyed when I told her about the temporary change of orders. We shipped our household belongings to Colorado and flew from Washington, DC, to Denver, Colorado. Denver, nicknamed "The Mile High City" because its official elevation is considered to be exactly one mile above sea level, is indeed a large city. This is not surprising, as Denver is a consolidated city and county. It is also the capital of the State of Colorado and the nineteenth most populous city in the United States.

Upon arrival in Denver, I immediately set out to find suitable housing for our family. We rented a spacious three-bedroom apartment near the Air Force base and settled in for our five-month stay in Denver. While I attended my classes at Lowry AFB, Vivian enrolled in undergraduate correspondence courses at a local community college.

Attending the Air Force Basic Intelligence Officer Course provided me with a solid foundation in many disciplines within the intelligence field that I would utilize later in my Air Force career. The course covered a wide range of topics, including intelligence gathering and analysis, threat assessments, information security, and various other tools and technologies. I received both classroom instruction and practical hands-on training in areas such as briefing, interrogation techniques, and intelligence operations planning. Overall, I felt a tremendous sense of fulfillment in being prepared for my future role in a highly classified environment where I would receive and interpret intelligence and possibly prepare briefs to support United States Air Force operations and America's international security initiatives.

In March of 1973, after completing the Basic Intelligence Officer Course, it was finally time for us to pack up our belongings and move to Offutt AFB to begin our initial tour of duty as Air Intelligence Officers with the 544th Air Intelligence Wing. However, Vivian informed me that she was pregnant with our third child before we left Denver. Naturally, no news could have made me happier at this stage. I considered myself the most fortunate of men. The Wallace Family was continuing to grow. My career was more than satisfactory since I was now a commissioned officer, an emerging Russian language specialist, and a qualified intelligence officer in the United States Air Force, all rolled into one. Most significantly, and perhaps very much in alignment with laying the foundation for a future happy marriage, my beautiful wife, with my unqualified support, was now on the threshold of her career aspirations in nursing. Our remarkable journey continued.

CHAPTER FIFTEEN

PRIVILEGED ASSIGNMENTS

"Do you see someone skilled in their work? They will serve before kings; they will not serve before officials of low rank."
- The Bible.

We finally arrived in Nebraska at the end of March 1973. After reporting for duty at the 544th Intelligence Analysis Wing, Strategic Air Command, Offutt AFB, my priority was to locate suitable living quarters for our family. As I mentioned earlier, we were expecting the birth of our third child by the end of December, and we needed more space for our growing family. I eventually rented a four-bedroom single-family house in the quaint, peaceful neighborhood of Bellview, Nebraska, less than a mile from Offutt. Vivian and I were quite pleased with our decision to live in Bellview. The community offered a small-town atmosphere while still being close to the amenities of the air base and downtown Omaha. In Bellview, there was a genuine sense of community where neighbors knew and cared about one another, and where streets and other common areas enjoyed high safety.

Due to my prior operational experience as an enlisted Non-Commissioned Officer (NCO) and Combat Aircrew Member, I was assigned to the position of Chief of the Soviet Air Forces Branch in the Directorate of Analysis and Production. This was quite an honor for a newly minted Second Lieutenant since the position had previously been held by an Air Force Captain. My reference to an early privileged assignment will be better appreciated when one considers that the rank of First Lieutenant still separated me from the rank of Air Force Captain. In June of 1973, I received a call from the secretary to the Director of Intelligence stating that the General would like to meet with me on a significant matter. Being relatively new to the organization, I had no idea what this meeting would entail. Keeping my mind open to all possibilities, I wasted no time reporting to the General's office. I saluted him smartly, and the General asked me to sit.

While staring intently at me, his next few words would throw me completely off balance.

"Lieutenant Wallace, I have observed your leadership and outstanding performance since you arrived at our organization. In two weeks, there will be a high-level meeting of members of the Intelligence Community in Washington, DC, to discuss Intelligence Requirements. I have selected you to attend this meeting as the representative of the Strategic Air Command." Needless to say, I was humbled by the General's positive comments about my work yet somewhat taken aback by the rare honor. I immediately asked myself, "Why would he propose sending *a junior officer to Washington to attend a high-level meeting with colonels and senior civilian executives?"* Nonetheless, I told the General that it would be an honor to attend this meeting and represent the best interests of the Strategic Air Command.

When I arrived in Washington, DC, for the meeting, I was ushered into a conference room with a large circular table. Representatives from many organizations in the intelligence community were already seated at the table. However, an empty chair was reserved for the Strategic Air Command (SAC) spokesman. As I attempted to walk up to the table to claim my seat, my escort pointed out that I was just a Second Lieutenant and instructed me to sit at the back of the room. Naturally, I complied with my escort's request. Just before the meeting commenced, the chairman asked who the representative of the Strategic Air Command was. From the back of the room, I stood up and replied that I was the Strategic Air Command representative. A puzzled look crossed his features, and the chairman requested that I take my seat at the conference table.

The surprised looks on the faces of the other participants at the meeting were not lost on me. One couldn't blame them. It wasn't every day that a United States Air Force combatant command sent a "lowly" Second Lieutenant to attend such a high-level meeting. After the two-hour meeting, the chairman thanked me for participating in the discussions. He acknowledged my expertise in addressing intelligence requirement issues of high interest to the Strategic Air Command. Upon returning to Offutt, I briefed the Director of Intelligence on the proceedings and results of the meeting. He stated that he was pleased with the outcome and thanked me for defending the best interests of our command at the forum.

Despite her earlier apprehension about living in Nebraska, Vivian's mood improved as soon as we settled into our new home in Bellview. On weekends, I would take the family on sightseeing trips to Omaha to visit popular attractions such as the Henry Doorly Zoo and Aquarium and the Old Market District, which features unique shops and restaurants. In addition to being a stay-at-home mom with two young boys, Vivian quickly established a network with Air Force wives living on base and within the Bellview community. She also had easy access to all base facilities, including the Base Hospital, where she scheduled her prenatal appointments. On the morning of December 31, 1973, the eve of the New Year, Vivian quietly told me it was time to go to the hospital for the birth of our third child. We drove the short distance from our home to the base hospital, and after a three-hour wait, our little girl was born. We named her Karen Nicole Wallace.

The following year of my tour at Offutt was uneventful until I received a call from my Air Force Assignments Officer in November 1974 informing me of another assignment change. He stated that the Air Force needed my expertise as an Intelligence Officer for a twelve-month tour of duty to support the Vietnam War effort as a member of the Strategic Air Command Advanced Echelon organization (SAC Advon) at Utapao Air Base in Thailand. SAC Advon personnel stationed at Utapao were responsible for overseeing and facilitating SAC operations in Southeast Asia, including bomber aircraft deployments, refueling operations, and other logistical and intelligence support functions. Having spent just over a year at Offutt with my family, I knew that the "needs of the Air Force" would always be a priority in my career, and I reluctantly accepted the assignment to Thailand. However, Vivian was disheartened by the news of my new assignment, as we had just gotten accustomed to our new life in Nebraska. Now, our entire family life was being uprooted. On a more positive note, since we lived in a rented house in Bellview, my wife and three children could remain in our residence until I returned from my twelve-month tour of duty in Southeast Asia.

Before departing for Thailand, I had to attend a two-month Intelligence Targeting Officer Course at Lowry AFB, Colorado. The Targeting Course was the first comprehensive course in the Air Force covering all aspects of *targeting* principles. A significant element of the course was that targeting is the key to successfully applying airpower, and graduates would form a solid core of trained targeting and weaponeering officers. We were taught that keen eyes are just as crucial on the ground as in the sky. These days, the United States Air Force has even specially trained targeting *analysts* who develop and implement tactical strike solutions for anything that poses a potential threat. Such experts analyze terrain and structures to make precise strike recommendations and battle damage assessments for their commanding officers. Targeting is a systematic, comprehensive, and continuous process that selects and prioritizes targets while matching the appropriate response with command objectives, operational requirements, and capabilities.

After completing the Targeting Officer Course in April 1974, I returned home to bid my wife and children farewell. Next, I scheduled a flight to Travis AFB in California to await instructions on my imminent flight to Thailand. However, to my great surprise, I received a message from the Air Force Assignments Officer informing me that my orders had been changed. Instead of going to Thailand as an Air Targeting Officer, my orders were changed to an assignment in Korea. The journey continues.

CHAPTER SIXTEEN

ON THE THRESHOLD OF HISTORY

"When I look into the future, it is so bright it burns my eyes."
- Oprah Winfrey

The flight from Los Angeles, California, to Seoul, Korea, lasted approximately twenty-two hours, with stops in Honolulu, Hawaii, and Tokyo, Japan. Upon arrival at Seoul International Airport, I was met by my Air Force sponsor, who drove me to Osan Air Base to begin processing for my twelve-month tour of duty at Kunsan Air Base. Located in Pyeongtaek, forty miles south of the capital city of Seoul, Osan Air Base is one of two major airfields operated by the U.S. in Korea, the other being Kunsan Air Base. Osan Air Base was the only United States Air Force (USAF) facility in the Republic of Korea that was entirely planned and built by the United States from scratch during the Korean War. The base has quite a fascinating history, mainly centered around what came to be known as the *Pueblo Crisis.*

The USS Pueblo was a United States Navy spy ship that gathered intelligence and oceanographic information and monitored electronic and radio signals from North Korea. On January 23, 1968, the ship was attacked and captured by a North Korean vessel in what became known as the "*Pueblo* incident" or the "*Pueblo* crisis." That incident precipitated the deployment of 1,000 USAF personnel were on temporary duty at Osan Air Base in support of Operation Combat Fox. Airmen from bases in the U.S. and Asia, including South Vietnam, arrived on January 25, within 48 hours of the attack.

The developing crisis underscored the importance of the installation at Osan. It improved existing facilities and constructed new structures, including aircraft shelters and control towers. Security was upgraded to support the increased tactical operations at the base, and from From January to March, over 6,500,000 pounds of cargo were shipped by rail to Osan. Although the *Pueblo* crisis subsided with the crew's release on December 23, 1968, fighter unit deployments continued to occur regularly. That response by the United States further increased fighter forces on the peninsula, eventually setting the stage for permanently assigned fighter units to return to South Korea. Osan Air Base would become home to various United States Air Force (USAF) units and support multiple missions, including air defense, reconnaissance, and combat operations. The base was strategically located to assist the region's American and South Korean military efforts.

After reporting for duty at the Air Force Element in Osan, I lobbied vigorously for a change in assignment to Osan due to its proximity to the vibrant city of Seoul. That effort, however, was to no avail as I was told that my expertise as an Air Targeting and Weapons Systems Officer was urgently required at the Eighth

Tactical Fighter Wing, Kunsan Air Base, Korea. On June 1, 1975, I departed Osan for the 125-mile, 3-hour bus ride to Kunsan Air Base. The airbase is on the country's western coast, near the Yellow Sea. Upon arrival at Kunsan, I reported to the Eighth Tactical Fighter Wing's Administration Office for processing and was assigned to the Officers' Quarters on base. Since this tour of duty was designated *"remote,"* dependents were prohibited from residing on the base.

In the Air Force, remote tours of duty refer to assignments where airmen are stationed in locations far from their permanent duty station, often in isolated or deployed environments, for an extended period, with such tours frequently lasting several months to a year or more. Naturally, remote tours are characterized by distance from family and friends, limited communication and connectivity, increased operational tempo, and potential for hostility and danger. As I would discover during the years of my exciting Air Force career, although remote tours can be challenging, they also offer opportunities for personal and professional growth and unique experiences. Airmen on remote tours often develop strong bonds with their fellow service members and gain valuable skills and perspectives.

After settling into my one-room quarters, I visited the Officers Club to introduce myself to some of my fellow airmen. I immediately spotted three Black Air Force First Lieutenants sitting at a table at the back of the club and went over to introduce myself to the group. Bill was the Chief of an Aircraft Maintenance Element, Ray was a Security Police Flight Commander, and CJ was a pilot flying the F4D Phantom fighter. After the round of introductions and pleasantries, I told the group I was the new Chief of the Air Targets Branch at Wing.

Headquarters. These three individuals would become very good friends and constant companions of mine. They showed me around the base and consistently offered sound advice on "surviving" at Kunsan as an African American Air Force Officer. To this day, I remain in contact with my "fellow black officers" from Kunsan, as we became lifelong friends, often reminiscing with pride and nostalgia about the time we spent together in Korea.

As Chief of the Air Targets Branch, a position previously held by a Captain, I supervised a ten-man team of enlisted personnel and junior officers whose primary mission was to assist our pilots and weapon systems officers in effectively applying air power. Specific components of our duties included target development, target analysis, force application planning, wargaming, mission planning, and target folder construction. Additionally, because of my previous training as a weapon systems officer, I frequently briefed the Wing Commander on selecting appropriate targets for strikes, developing weapons and force requirements, and managing targeting resources. Later in my Air Force career, my close association with Colonel Charles Hamm, the Eighth Tactical Fighter Wing Commander, would prove fortuitous. He was a former member of the Air Force Air Demonstration Squadron, the *Thunderbirds*, later became the Superintendent of the United States Air Force (USAF) Academy, and retired as a three-star Lieutenant General in the United States Air Force. I will elaborate more on General Hamm in later chapters.

Although I excelled as the Wing Air Targets Officer, with only four months remaining on my tour of duty, I received some troubling news. The squadron commander called me into his office and informed me that my bank had returned a check I had written at the Base Exchange due to "insufficient funds."

Because it was Air Force policy that all officers maintain financial responsibility at all times, the commander soberly conveyed that this singular incident could earn me a written reprimand or a possible dismissal from the Air Force. My first reaction was one of total disbelief. Before departing for Korea, I had taken extra steps to ensure that our family's financial resources were secure. I thanked the commander for conveying this information and assured him that I would rectify the problem immediately.

After numerous attempts to contact Vivian in Nebraska by phone from Korea, I finally spoke with her. Our conversation was incredibly awkward. *"Hello darling, this is Lewis calling you from Korea."* Her response was chilling.

"Lewis, who?"

At that moment, I knew something was wrong. I pleaded with my wife not to write any more checks on our bank account until the end of the month because I had just received a returned check notice from the base exchange. Reluctantly, she agreed to take my advice. Yet, I remained stunned by the strange tone in her voice and the oddity of our brief exchange. Later, I would find out that my wife was suffering from acute bouts of depression and anxiety and was self-medicating with opioids. Needless to say, this was very troubling news. Thankfully, I got immediate relief from my check issue, as three officer friends at Kunsan loaned me the money to cover the insufficient funds, saving my promising Air Force career.

Three months before my tour at Kunsan ended, I contacted the Officer Assignments Office again to advocate for my next assignment. Due to my urgent family situation, I requested to be assigned to an intelligence position in the Washington, DC, area, where I could obtain proper medical care for Vivian. To my utter amazement, the Assignments Branch rejected my request. I was informed that I had been selected to attend a one-year tour at Ohio State University under the Air Force Institute of Technology's Civilian Institutions Graduate School Program. After graduating with a Master of Arts degree in Slavic Languages and Literature and Foreign Language Education, I would be assigned to the USAF Academy as an instructor in Russian Language. Naturally, I was overjoyed at this news and readily accepted the new assignment. After years of studying the Russian language and culture, I finally began to see the fruits of my labor in the most rewarding ways, especially since I would also be making history. My career trajectory indicated that I would eventually work at the USAF Academy as the first African American instructor in Russian in the history of this prestigious United States Air Force institution. The horizon couldn't look any brighter.

CHAPTER SEVENTEEN

THE DISTINGUISHED GRADUATE

"You live a new life for every new language you speak. If you know only one language, you live only once."
- Unknown Author

In May 1975, I departed Kunsan Air Base, Korea, for a long flight back to Nebraska. Upon arriving at Omaha International Airport, I spotted Vivian and my three children waiting in the reception hall, waving a giant "Welcome Home Dad" card. Vivian was also carrying a large bouquet of red roses. I was overwhelmed by this special homecoming with my family. Although, as an Air Force officer, I understood that it was sometimes necessary, if not an unavoidable obligation, to answer the "call to duty" and serve in remote regions of the world, being separated from my loved ones was never easy. Returning home and being warmly greeted by my family was always a source of immense joy.

Life in the nation's armed services is a unique experience for service members and their families. I belong to those who sincerely believe that our service men and women in the Armed

Forces deserve a special place in our hearts for the extraordinary sacrifice of giving up the comforts of domestic family life, especially for the precarious missions they must undertake outside the United States. Of course, not all military deployments are of equal magnitude in terms of that sacrifice. However, regardless of whether short or long, they have one thing in common: separation from family.

Global conflict and unrest have led to the deployment of large numbers of military personnel, whether active duty, Reserves, or National Guard. As a result of duty assignments, military members are often separated from their families for lengthy periods and sent to distant, dangerous, or unknown locations. A family that loses the active presence of a spouse and parent through separation will face significant challenges and stress. During such a deployment, family members may feel isolated, unsupported, and anxious. They may also experience financial stress. Media coverage of global conflicts can further heighten concerns. Some families must also cope with the trauma of having a service member seriously injured or killed. In families with existing medical or emotional challenges, having a member away can be especially difficult. I take this moment to solemnly pay tribute to the gallant men and women who have committed their lives to the protection and preservation of our dear nation.

After packing up our household goods in Bellview, Nebraska, for the trip to Columbus, Ohio, we drove back to Washington, DC, for a much-needed three-week vacation to visit with family and friends. The time spent in Washington allowed me to reconnect with Vivian and discuss some serious issues she faced while I was in Korea. We agreed that she needed medical treatment and counseling for her apparent addiction to opioids. I promised her that as soon as we arrived in Columbus, I would seek the best medical resources available so she could overcome both her affliction and addiction.

After we arrived in Columbus, our first stop was to check in at Rickenbacker Air Force Base, located approximately ten miles southeast of the city. Rickenbacker AFB has a fascinating history. During World War II, the installation served as a US Army Air Forces training base known as Lockbourne Army Airfield. It became an Air Force base in 1948, shortly after the establishment of the United States Air Force as an independent branch of the US armed forces. The base was named Lockbourne AFB from 1948 to 1974, and later Rickenbacker AFB in 1974.

I had previously requested a four-bedroom townhouse on Rickenbacker Air Force Base. My request was granted, and this would be home for the family for the next twelve months while I attended graduate school at Ohio State University. Ohio State University, founded in 1870, is a major public research university with a diverse student population and a wide range of academic programs. My objective was to graduate in a year by completing a Master of Arts degree in Slavic Languages and Literature and Foreign Language Education. However, since I had not taken any academic courses since graduating from Syracuse University in 1972, the thought of completing such a rigorous course of study in just one year initially seemed like a Herculean, if not impossible, task. Yet again, I recalled my mother's words, who always told me that with perseverance and hard work, you can accomplish anything in life.

For the benefit of the uninitiated, the Slavic languages are quite fascinating. For over a thousand years of recorded history, the places and peoples of today's Eastern Europe and Russia have generated considerable curiosity from all over the world. Central to these peoples and cultures are the Slavic languages: Russian, Ukrainian, and Belarusian to the east; Polish, Czech, and Slovak to the west; and Slovenian, Bosnian/Croatian/Serbian, Macedonian, and Bulgarian to the south. The Slavic languages are closely related to each other but are also related to the Germanic languages, including English and others in the Indo-European family. Despite the linguistic similarities among the Slavic languages, these countries and peoples have followed different paths in culture, religion, history, and political tradition. However, these paths have frequently crossed in the creation and disintegration of empires in Eastern Europe's constantly changing political landscape.

My mother's inspiring words motivated me to buckle down and complete my courses. At graduation in June 1976, I attained a 3.5 GPA in all my courses and was designated a *"Distinguished Graduate"* by the Air Force Institute of Technology's Civilian Institutions Graduate School Program. Following my graduation from Ohio State, my next assignment was to the United States Air Force Academy as the first African American Instructor of Russian. I requested that the Air Force send me to Moscow for additional training in teaching the Russian language to foreigners before arriving at the Academy. To my surprise, the Air Force granted my request and authorized me to travel to Moscow with fifteen American graduate students for a two-month study tour at the Pushkin Foreign Language Institute.

The Pushkin Institute is part of Moscow State University, specializing in teaching Russian as a foreign language. It is one of the most prestigious language institutes in Russia and offers a wide range of courses and programs in Russian language and literature for international students from various countries.

During my time at the Pushkin Institute, I had the opportunity to immerse myself in the Russian language and culture in the heart of Moscow and gain a deeper understanding of the country and its people. Naturally, it was difficult for me to leave Vivian and the kids alone in Columbus again while studying in Russia. I was distraught about her previous bouts of anxiety and depression, and I hoped she would continue to improve in health while I was away. She was a strong woman whom I loved very much. However, in my opinion, this was a once-in-a-lifetime opportunity to travel to Russia to enhance my teaching credentials as a new Russian language instructor. At the end of June 1976, I began the long trip by airplane and train from Washington, DC, to Moscow, Russia. My incredible journey continues.

CHAPTER EIGHTEEN

WELCOME TO MOSCOW!

"You can always tell a man's nationality by introducing him to a beautiful girl. An Englishman shakes her hand; a Frenchman kisses her hand; an American asks her for a date; and a Russian wires Moscow for instructions."
- Anna Karina

It all felt like a dream. I had studied the Russian language for so many years that I had virtually become a *Russian* living on American soil. But now, I was finally going to Russia, not only to speak the Russian language on Russian soil but also to learn how to teach it to others. After contacting a Danish tour agency, I joined a team of fourteen other U.S. graduate students going on a two-month foreign exchange tour at the Pushkin State Russian Language Institute in Moscow.

The Pushkin State Russian Language Institute is a leading Russian language education and research institution in Moscow. Founded in 1966 and named after Alexander Pushkin, the great Russian poet and writer of African ancestry through his great-grandfather, Pushkin is considered the founder of modern Russian literature and is revered as Russia's national poet. The mission of the Pushkin Institute is to promote Russian language and culture worldwide. Well-established and highly regarded for its uncompromising academic standards and intensive language programs in Russian and other foreign languages, it offers courses for foreigners, teacher training, linguistics and philology research, and cultural exchange programs. With an alumni base of over 50,000 graduates from around the world, the Pushkin Institute, a renowned center for Russian language and culture, attracts students, scholars, and professionals globally. Its programs and research focus on promoting the Russian language and culture while fostering international understanding and cooperation.

The plan involved all members of our exchange group flying to West Berlin, Germany, where, after a brief rendezvous, the group was scheduled to travel by train from Berlin to Moscow. However, the circumstances in international politics and relations were somewhat complicated, and it should have been a simple and seamless trip. In the summer of 1977, the world was in the midst of the *Cold War era*. All travelers between West Berlin and East Berlin were required to pass through checkpoints operated by East German border guards. A rather amusing incident occurred when our study group arrived at the East German checkpoint. While the fourteen other members of our group were processed quickly, I was instructed by the border guard to exit the line for further questioning. Although I presented my American passport and my entry visa into the Soviet Union, the border guard continued to question me about why, as a Black man, I was trying to "defect" to Russia.

Finally, the tour leader personally intervened on my behalf, stating that I was a member of the authorized study group. After many years of reflection, I still do not know if my experience was common for any Black man traveling to Moscow or if there was some other ulterior motive in detaining me, a US Air Force officer, at an East German checkpoint. The stark differences between the two Berlins became readily apparent when we emerged from the checkpoint and entered East Berlin. West Berlin was a vibrant city where citizens enjoyed a high standard of living, tremendous personal freedoms, and access to consumer goods. In sharp contrast, the citizens of East Berlin had limited personal freedoms and were subject to strict control over all aspects of their lives. Specifically, East Berliners faced travel restrictions, limited access to Western media and goods, and constant surveillance by the government's secret police. Indeed, the division between East and West Berlin was a physical and ideological representation of the *Cold War,* with the Berlin Wall serving as a powerful symbol of separation and oppression. The reunification of Germany in 1990 marked the end of this division, and Berlin has since become a thriving, unified city.

After a two-hour wait at a dimly lit East German train station, we finally departed on our thirty-hour journey to Moscow. A peculiar feature of the train coaches was that all their windows were firmly locked and covered with black paint. So, during the entire trip, we could not view the passing countryside. We finally arrived at Leningradsky Railway Station in Moscow on a bright Saturday morning. After retrieving our luggage, we boarded a bus to a restaurant for breakfast. My first meal in Moscow was disgusting, to say the least. After traveling by train for nearly thirty hours from Berlin to Moscow, the runny eggs, stale black bread, and weak tea served to our group for breakfast caused most of us serious gastrointestinal problems. *Welcome to Moscow!*

After breakfast, instead of proceeding directly to our rooms at the University Hotel, adjacent to Moscow State University, the bus stopped at a construction site near the Pushkin Institute, where we were told we had "volunteered" as construction workers to help build a new dormitory for the Pushkin Institute. During the Soviet era, there was a tradition called *"Subbotnik."* The word was derived from the Russian name for Saturday, *"subbota."* The Subbotnik was a day dedicated to voluntary community service, and that day was usually a Saturday. Although participation in these Saturday community service activities was often mandatory for workers, for a group of foreign exchange students on their first day in Moscow, helping build a new dormitory was an incredible and rewarding experience. After completing our obligatory community service, the bus took us to the University Hotel, adjacent to Moscow State University. The hotel, managed by the Soviet Ministry of Education, catered to a wide range of guests, including international students, academic travelers, and special guests. Russian standards rated our two-student room accommodation quite spacious, mainly because many Russian graduate students lived in small, crowded dormitory rooms with four or five roommates.

The curriculum at the Pushkin Institute included courses in grammar, vocabulary, reading, writing, speaking, and listening comprehension, and foreign language teaching methodology. Classes were held from 9:00 AM to 4:00 PM, Monday through Friday, with a one-hour break for lunch. Because the Soviet Ministry of Education sanctioned our exchange group, we were not allowed to eat lunch at the regular student dining facilities at

Moscow State University, where the cost of a bowl of soup or a sandwich was about one ruble. Instead, during the lunch break, we were bused to the Central House of Writers Restaurant, which was closed to ordinary Russian citizens. At the restaurant, we could purchase an excellent four-course meal for just one ruble. The disparity in what one ruble could buy in Moscow for an ordinary Russian citizen versus someone from the privileged elite of Communist society was eye-opening.

One benefit of being a foreign exchange student in Moscow was receiving an official student identification (ID) card. With my student ID card, I was eligible for discounts on transportation, museums, attractions, and other cultural experiences in the city. Carrying an official student ID also served as proof of identity if stopped by the police or other security personnel. Additionally, many universities in Moscow provided access to their facilities, such as libraries, sports complexes, and study spaces, only to students with valid ID cards. After my daily classes at the Pushkin Institute, I often ventured out alone to explore some of the many sightseeing attractions in Moscow. Each site offered a unique glimpse into the vibrant city's history, culture, and beauty. Some of the most popular sightseeing sites in Moscow are Red Square, the Bolshoi Theater, the Moscow Metro, the Pushkin Museum of Fine Arts, Gorky Park, the Exhibition of Achievements of the Soviet National Economy (VDNKh) complex, and the Arbat pedestrian shopping street.

To improve my Russian language proficiency and to better understand what the ordinary Russian citizen thought of the state of relations between Russia and the West, I frequently sat on a park bench and talked with strangers. Some Russians, who likely held negative stereotypes or prejudices against Black people, rebuffed my efforts to communicate with them, believing that I was African. However, on other occasions, as soon as I introduced myself as an African American, their perception of me changed to one of curiosity and even warmth.

One day, I spoke with an elderly Russian gentleman in a park. When I told him that I was American, he was astonished.

"In our country, the newspapers always say that Blacks in America are criminals who live in poverty and are constantly persecuted and discriminated against by their government," he said, staring at me incredulously.

"Look at me," I replied, smiling. *"I am an African American man who is not a criminal. I have money, am healthy, well-fed, and dressed in the latest blue jeans and athletic shoes."*

"Then why does our government keep lying to us about the state of Blacks in America?" he asked, a perplexed look on his face.

Unfortunately, I was in no position to offer an appropriate response to his question. I suggested he address the question to the local Soviet authorities.

During my solo wanderings on the streets of Moscow, I never felt that my safety was in jeopardy. Of course, there were the usual curious stares and the occasional racial slurs directed at me as a Black man. However, while studying in Moscow, I had the opportunity to meet several groups of African students studying at both Moscow State University and Friendship University, who told me they faced significant racial discrimination and prejudice daily. These students often encountered discriminatory practices, including difficulties finding housing, incidents of verbal abuse, and limited access to resources and opportunities. Additionally, African students were frequently targets of assault and harassment by both Russian "Skinhead Gangs" and the local police.

Additionally, the students said that they were subjected to racial slurs and segregation in public spaces, such as public transportation and restaurants. I recall trying to enter a popular restaurant in downtown Moscow but was abruptly stopped by the doorman, who said there were no tables available for me, a Black man. However, as soon as I told him I was from America, a capitalist country, his attitude towards me changed instantly, and I was granted entrance to the restaurant.

A highlight of that first trip to Moscow in 1977 was meeting two remarkable individuals: a Black Soviet journalist and an African American who had immigrated to Russia during the Great Depression. Yelena Khanga, a Black Russian, was a prominent journalist, television producer, and author born in Moscow to a Russian mother and a Tanzanian father. She grew up in a multicultural household in Russia, which influenced her perspective on race, identity, and diversity. George Tynes, on the other hand, was a native of Norfolk, Virginia, and a graduate of Wilberforce University. He emigrated to the Soviet Union in 1933. As an agricultural specialist, George was sent to various Soviet republics to teach people how to raise ducks and other waterfowl and became a nationally recognized expert on poultry.

Although he never joined the Communist Party, George became a Soviet citizen, married a Russian woman, and had three children. After his retirement, George received the *"Hero of Socialist Labor Medal,"* awarded to individuals who made outstanding contributions to the advancement of society and the Soviet state. He also received a spacious two-bedroom subsidized apartment in Moscow. When I visited George at his residence in 1977, his first question was, *"What is the current situation of Blacks in America?"* He also asked me to bring him some copies of Black periodicals, such as *Ebony* and *Jet* magazines, and some of the latest jazz records whenever I returned to Moscow. Regrettably, George died at his home in Moscow in 1982.

On the day of my departure from Moscow, I had an extremely traumatic experience. When I arrived at the customs control point at Sheremetyevo International Airport, the customs inspector asked me to step out of line for further questioning. He led me to a small interrogation room where my person and baggage were thoroughly searched. He then reached into one of my suitcases and removed five personal letters given to me by individual US embassy employees to be mailed when I returned to the United States. The customs officer confiscated the letters and harshly admonished me for attempting to take them out of Russia without prior approval. Dumbfounded, I claimed ignorance of Soviet laws and was finally released to catch my flight home. In retrospect, it seemed rather curious that, on both my entry and departure from Russia, I was singled out for "special treatment" by the authorities. *Was it because I was an African American on an exchange visit to Russia? Was it because I was a United States Air Force Officer?* In all probability, I will never know.

CHAPTER NINETEEN

THE SINGLE PARENT TEACHER

"When the going gets tough, the tough get going."
- Joseph P. Kennedy (1888-1969). United States Ambassador to the United Kingdom and father of the 35th President of the United States, John F. Kennedy.

During the three-hour flight from Moscow to Frankfurt, Germany, I had ample time to reflect on my first trip to Russia. Overall, the two-month study tour to Moscow went exceedingly well, and it afforded me, an African American, the rare opportunity to experience the Russian language and culture firsthand. Immediately after clearing customs at Frankfurt Airport, I called Vivian to tell her when I would arrive at Columbus Airport. To my surprise, I did not receive a warm reception from my wife upon my return from this overseas tour. She merely said she would see me when I got home. When I finally landed at Columbus Airport, no one awaited me in the reception hall. There were no banners or flowers for me this time. Vivian and the kids did not even come to the airport to welcome me back, and sadly, I had to take a taxi to our residence on Rickenbacker AFB.

After warmly greeting Vivian and the kids when I arrived home, I decided to talk seriously with my wife to determine why she was acting so strangely. During our conversation, Vivian told me that while I was in Russia, she'd had a relapse. Her bouts of depression and anxiety had worsened, and she was once more taking opioids for her condition. Additionally, Vivian told me she wanted a legal separation and did not want to join me at my new duty location at the United States Air Force Academy (USAFA) in Colorado. Instead, she wished to remain in Columbus with our three children. I felt as if I had been hit with a ton of bricks. Vivian's statement was a punch to the gut, as the emotional impact of her sudden and unexpected request left me stunned and overwhelmed. Quickly regaining my composure, I repeatedly attempted to change Vivian's mind, but it was all to no avail. Finally, we reached a mutual agreement. We settled on a legal separation that would jeopardize our eleven-year marriage. Given all we had been through together, the pain of leaving Ohio without my family was both disappointing and heartbreaking.

My orders required that I arrive at the Air Force Academy in Colorado Springs by early August to complete the new instructor faculty orientation program before classes began in September. After renting a three-bedroom house for Vivian and the kids in downtown Columbus, I contacted a realtor in Colorado Springs and rented a two-bedroom apartment about ten miles from the Academy. Finally, I loaded my car and began the 1,200-mile drive from Columbus to Colorado Springs. When I arrived in Colorado Springs in August of 1977, the city was undergoing a period of accelerated growth and development. Located at the base of Pikes Peak, which rises some 14,000 feet above sea level on the eastern edge of the Southern Rocky Mountains, Colorado Springs, the county seat of El Paso County, Colorado, was known for its stunning natural beauty and outdoor recreational activities. The city was also home to a diverse population that included military personnel, retirees, and families attracted to the area for its quality of life. Colorado Springs's economy has always been driven primarily by the military and tourism, in that order.

The altitude of Colorado Springs is approximately 6,035 feet above sea level. Since I was not accustomed to living at such heights, I began experiencing the typical symptoms of altitude sickness, including shortness of breath, fatigue, cluster headaches, and dehydration, soon after my arrival. It took me almost a week to acclimate to Colorado Springs' high altitude. Then, I reported to the Academy to begin my three-year tour of duty as a Russian language instructor.

The United States Air Force Academy is an institution dedicated to training and educating future officers in the United States Air Force, equipping them intellectually and militarily to contribute significantly to the defense of our country. It was founded in 1954 in Colorado Springs, Colorado. The curriculum of the USAFA focuses on academics, military training, athletics, and character development. Upon graduation, cadets receive a Bachelor of Science degree and are commissioned as Second Lieutenants in the United States Air Force. The academy prepares cadets for service in the officer corps of the United States Air Force. Although it is the youngest of the five service academies, having graduated its first class in 1959, it is the third in seniority.

The academy was also one of Colorado's most prominent tourist attractions, attracting thousands of visitors yearly. Upon entering the faculty building for the first time, I was overwhelmed with a sense of satisfaction and immense pride. Out of hundreds of other Air Force Officers who applied for the position of Instructor of Russian, the Chairman of the Department of Foreign Languages had selected me, an African American First Lieutenant, for the job. Additionally, he obtained approval from the Air Force Personnel Center to sponsor my advanced degree at Ohio State University and fund my follow-on exchange study tour to Moscow.

When I arrived at the Air Force Academy in the summer of 1977, the Department of Foreign Languages taught cadets various languages. Some of the languages offered included Arabic, Chinese, French, German, Russian, and Spanish. Cadets had the opportunity to learn these languages through multiple courses and programs designed to develop their linguistic proficiency and cultural understanding. Although most cadets opted to fulfill their foreign language requirements by enrolling in traditional French, German, and Spanish courses, many students were interested in learning the so-called "strategic languages," which included Arabic, Chinese, and Russian.

After meeting with the Department Chairman, I was escorted to my new office. To my amazement, even as a new instructor, I was given a spacious corner office furnished with neomodern taste and a commanding view of the Colorado mountains and the beautiful Academy Chapel, an iconic symbol of the academy campus. This office was beyond my wildest dreams. Later, I introduced myself to the three other Russian instructors in our section. Although these individuals held the military ranks of Captain, Major, or Lt Colonel, had excellent academic credentials, and were fluent in

Russian through their previous Air Force assignments, none had ever personally visited the Soviet Union. My fellow instructors complimented my excellent teaching qualifications, stating that my addition to the faculty as a Russian instructor would be a valuable asset to the Foreign Languages Department because I could convey, in great detail, my lived experiences while living and studying in the Soviet Union.

As the "new kid on the block," I was assigned to teach four of the six basic Russian classes. Because I was a new instructor, my first teaching class began at 8:00 AM, three days per week. Since I had arrived in Colorado Springs without my family, the early teaching schedule was initially of no great consequence. However, later in this chapter, it will become apparent how this early teaching schedule adversely affected my entire tenure at the Academy. In addition to my teaching load, I had additional duties as the Basic Russian Course Director. In that capacity, I was responsible for program administration, course development, assessment and evaluation, and student support for more than one hundred cadets enrolled in the basic Russian course.

An amusing incident occurred on my first day teaching Russian at the Academy. According to protocol, when an instructor entered a classroom, all cadets were required to stand at attention to greet the instructor. As I entered the classroom on that first day, the cadet class leader, looked dumbfounded at me and jokingly asked, *"Sir, is this the Swahili class or the Russian class?"* Without losing my composure, I told the student, first in Russian and then in English, that this was the basic Russian class. After that firm yet civil response, there were no more questions about which course the cadets were taking.

Although new cadets admitted to the Air Force Academy typically have strong academic qualifications, including robust performance in high school in the core subjects of mathematics, science, and English, I was amazed to find that most cadets in my basic Russian classes were deficient in English grammar. Consequently, my initial challenge was to review their understanding of the basic concepts of English grammar before attempting to teach them the Russian alphabet, basic vocabulary, and grammar concepts. Through patience, perseverance, and hard work, I successfully taught my students the Cyrillic alphabet and basic listening, reading, writing, and speaking skills in Russian by the end of the second semester. Additionally, I conducted weekly lectures on Russian history and culture and shared with my students how ordinary Russian people survived daily under the oppressive Soviet Communist system.

Despite achieving great success in my duties as a Russian instructor at the Academy, my personal life was filled with heartache and despair. I missed my wife and kids dearly, especially during the holidays, and constantly attempted to convince Vivian to relocate to Colorado Springs so we could be together as a family again. Unfortunately, my pleading continued to fall on deaf ears. Finally, in January of 1978, I moved out of my apartment and purchased my first home in Colorado Springs. Somehow, I entertained the notion that if I bought a house in a friendly residential community near the academy, Vivian would eventually relent and agree to join me. Sadly, that notion was not only misguided, but it also turned out to be a forlorn hope.

One day, while teaching my Russian class, I received an urgent call from an attorney in Columbus. He informed me that Vivian had been involved in an accident in which a pedestrian was fatally injured. Additionally, Vivian had been charged with vehicular homicide and taken into custody. Initially, it was difficult for me to process this information. I asked the attorney about the status of my three children, and he said they were safe and temporarily staying with neighbors. I thanked the attorney for the call and told him I would fly to Columbus the next day to arrange a bond for Vivian and take care of my kids. Due to the seriousness of the situation, I immediately requested approval from my supervisor for emergency leave to travel to Columbus and attend to my family.

Upon my arrival in Columbus, my first order of business was to arrange bail for Vivian so she could be released from jail. Then, I collected my kids from the neighbor's house and moved them back into Vivian's apartment. When I opened the door to the apartment, I couldn't believe my eyes. The entire space was in a shocking state of filth, with unmade beds, trash, and clothing strewn across almost every room. It was evident that Vivian's mental decline due to her drug addiction had worsened, and she was now unable to care for our children properly. Steven, my oldest son, told me he was doing everything he could to care for his brother and sister because their mom was ill most of the time. I immediately reserved two rooms at a local hotel, packed up some of the kids' belongings, and temporarily left the apartment.

Luckily, I posted a bond for Vivian the next day, and she was released from jail. Although she was distraught and highly remorseful about the accident and the death of the pedestrian, she told me that she had been late for work and, in her hurry, lost control of the car at an intersection. After comforting her about the traumatic experience, I suggested that it would be best if she went home to stay with her parents in Beltsville, Maryland, until the judicial proceedings were completed. In the meantime, I would take the kids with me back to Colorado Springs. Finally, she agreed that this would be the best course of action for everyone involved. Several days later, after Vivian was safely home with her parents, I rented a U-Haul truck, loaded the kids and some of their belongings, and made the 1,200-mile trip back to Colorado.

On arriving back at my home in Colorado Springs, I was filled with both joy and sadness. On one hand, I reunited with Steven, Marten, and Karen. On the other hand, I felt a deep sense of despair and anxiety at the thought of being permanently separated from my wife after almost twelve years of marriage. Another concern weighing heavily on my mind was how, as a single parent, I could successfully raise my three children, aged eleven, nine, and five, with little or no help. My mother, quite expectedly, offered to temporarily take in the children while I worked at the academy, but I politely rejected her offer. Despite all odds, I believed it was my responsibility to keep my children together and with me.

Becoming a single parent drastically changed my life. I faced five principal challenges. I had to exert a near-Herculean effort to balance work and parenting responsibilities. I struggled to provide adequate emotional support for my children. I experienced financial strain due to additional expenses for childcare and extracurricular activities. I dealt with feelings of isolation and loneliness. Finally, I coped with high levels of stress and pressure because of fewer opportunities to socialize with friends in Colorado Springs.

To complicate an already challenging situation, although I was now a single parent with three children to care for, I did not receive much consideration or support from my superiors in the Foreign Languages Department. I was still required to teach my first Russian class at 08:00 am while managing the daunting daily task of preparing two of my sons for school and dropping off my daughter at a babysitter, all before reporting to class. To make matters worse, the babysitter could only care for my daughter until noon. As a result, I had to use my lunch period to pick up my daughter from one babysitter and deliver her to another. Sadly, I did not have time to eat lunch before my next Russian class began at 1:00 pm, and hunger pangs became my constant companion until the close of classes each day.

That period was exceptionally challenging for me. As the saying goes, *"When the going gets tough, the tough get going."* This means that strong individuals will work harder to meet challenges when situations become complicated. It was also a time for me to discover that what lies behind us and what lies ahead of us are irrelevant compared to what lies within us. This mindset helped me keep my situation in perspective. Indeed, I could see that what appeared so difficult was minor compared to my determination. In the final analysis, Dr. Martin Luther King Jr. once said, *"The ultimate measure of a man is not where he stands in moments of comfort and convenience, but where he stands at times of challenge and controversy."*

Over the next month, I managed to balance the needs and schedules of my three children while effectively prioritizing tasks and activities at home and work. Although the kids missed their mother, the situation couldn't be helped, and she had no further communication with them for over six months. The news arrived that the judge in Columbus had acquitted her of all charges concerning the fatal accident, and she was now a free woman. When I asked Vivian to join us in Colorado to become a complete family again, she refused my invitation. As matters stood, I was left with no option but to file for divorce. Since Vivian refused to attend the divorce proceedings in person, the judge ordered that I receive sole custody of our three children.

Despite the demanding obligations of being a single parent, I continued to excel in my teaching duties. I received high marks from both my students and supervisors for my outstanding performance. However, I was confronted with a serious and confounding dilemma. According to Academy regulations, a military officer who served satisfactorily as an instructor and course director for two years was automatically promoted to the academic rank of *Assistant Professor*. To my utter astonishment, when I asked my supervisor about my academic promotion, he coldly replied that at a meeting of the leadership of the Foreign Languages Department, it was decided that they would not support my promotion to Assistant Professor of Russian. When I asked why my promotion was declined, I did not receive a response from my supervisor. Finally, feeling disappointed and hurt about how I was being treated, I informed my supervisor that if I did not receive the promotion I rightly deserved, I would respectfully request an immediate reassignment from the Air Force Academy. After about an hour of closed-door deliberations, my supervisor exited the meeting, shook my hand, and welcomed me as the new Assistant Professor of Russian. I was delighted by this turn of events. Becoming the first African American in the twenty-five-year history of the Academy's Foreign Languages Department was a distinct honor, and I felt a particular sense of genuine fulfillment.

During my final year at the Academy, I was tasked with planning and leading the first-ever study tour of the Soviet Union by Air Force Academy cadets. After a series of delicate negotiations between the Department of Defense and the Soviet Ministry of

Defense, I finally obtained approval to accompany a group of thirteen cadets to the Soviet Union in the spring. Before departing for Russia, I met with the Academy's Commandant of Cadets, a Brigadier General, who told me, in no uncertain terms, that if I did not return to the Academy with all of his cadets, I could just as well remain in Russia. Of course, I heeded the General's advice.

The first study tour by the United States Air Force Academy cadets to the Soviet Union was a tremendous success. It exposed the cadets to Russian culture, history, and military practices. During the two-week trip, they toured military installations, met with their Soviet counterparts, and engaged in cultural exchanges with the local population. Overall, the visit was a resounding success and was seen not only as a significant step toward improving US-Soviet relations in the Cold War era but also as a means of fostering mutual respect and understanding despite the political tensions of the time.

During my last three months at the academy, I had another disappointing experience. At a faculty meeting, the Department Chairman stated that he had received authorization from the Air Force to sponsor a faculty member, likely a minority, to complete a doctorate and then return to the Academy as a tenured professor. Being the only African American in the department, I had high hopes of being selected to return to school. However, this was not the case. To my utter disappointment, the Department Chairman selected a Caucasian instructor who taught Spanish for this prestigious assignment.

In May of 1980, I received an unexpected call from my former wife, Vivian. During our conversation, and to my surprise, she informed me that she planned to move to Colorado Springs soon due to health reasons. According to Vivian, she was now fully recovered from her addiction to opioids. However, doctors had diagnosed her with severe respiratory problems, including a chronic case of asthma. Her doctors had recommended that moving to a mountainous state like Colorado, with its high altitude and dry climate, might help alleviate her condition. I was saddened to hear the news of Vivian's medical problems and told her that, although the kids and I were leaving for Washington, DC, by the end of June, I would gladly assist in facilitating her move to Colorado.

After eleven years of marriage and three kids, I still loved this woman, even though we were now divorced. I told Vivian our children would be thrilled to spend time with their mother again and that I would help her settle in after arriving in Colorado Springs. Once Vivian was in town, I assisted her in renting an apartment, buying a used car, and finding a job. Before I left for Washington, DC, I had several conversations with Vivian to see if reconciliation was possible, even if only for our children's well-being and stability.

However, to my surprise, she rejected my efforts at reconciliation and ultimately told me that, in her opinion, our children would be better off living with me. Although her response disheartened me, I told Vivian that I wished her well and hoped she would stay in touch with me and the kids in the future.

Undeterred by this unfortunate turn of events, I completed my tour of duty at the United States Air Force Academy with distinction and left Colorado Springs for my follow-on assignment to Washington, DC, as a Russian Area Specialist at the Department of Defense. Returning to the nation's capital would allow me, with the support of my family and friends, to establish a strong safety net for my three children while continuing to pursue my dream of becoming a US Air Force Attaché at the American Embassy in Moscow. My incredible journey continues.

Estimates presented their own unique challenges.

I had to balance my responsibilities at work with my duties at home, ensuring that my children were well cared for and that I was meeting the expectations of my job. The support services available on Bolling AFB were invaluable, providing me with resources that helped ease some of the burdens of single parenthood. I relied heavily on the base's daycare facilities, which allowed me to focus on my work while knowing my children were in safe hands.

Despite the challenges, I embraced the opportunity to grow professionally and personally. My role as an Air Intelligence Analyst was demanding, but it was also rewarding. I found myself immersed in complex intelligence assessments, collaborating with experienced professionals, and contributing to significant national security decisions. This experience not only enhanced my analytical skills but also deepened my understanding of the intricacies of defense intelligence.

As I navigated this new chapter in my life, I remained committed to my family. I made it a priority to spend quality time with my children, instilling in them the values of hard work, resilience, and the importance of education. We often explored Washington, DC, visiting museums, parks, and historical landmarks, creating lasting memories together.

In the midst of my professional journey, I discovered that love and support could be found in unexpected places. The bond with my children grew stronger as we faced challenges together. I learned to appreciate the little moments—family dinners, bedtime stories, and weekend outings—reminding me that love truly is the greatest strength of all.

Estimates of the Defense Intelligence Agency proved to be a genuinely daunting task. However, with God's grace, the continued support of my mother and other family members, and the eventual engagement of an excellent housekeeper for the kids, I was able to successfully juggle my family and work obligations. Although I was not selected to return to school for my PhD in Russian while at the Air Force Academy, I never gave up on my dream of obtaining my doctorate. In moments of disappointment, I continued to recall the words of my mother, who, instilling in me the power of positive thinking, had constantly told me that through perseverance and hard work, I could achieve all my life goals. Also, an adage that consistently remained at the back of my mind was, *"When one door closes, another one opens."*

Finally, after three months at the Defense Intelligence Agency, I applied for and was accepted as a student in Georgetown University's very competitive Doctoral Program in Russian Area Studies. In my admissions essay, I proudly pointed out that my mother had worked for several years as a housekeeper and maid for priests who resided at the rectory at Georgetown University. My mother had always been blessed with an incomparable ability to predict the future, if not even with an extraordinary form of prescience. The consummate visionary, she would always tell people that someday her son, an African American, would be a doctoral student at this prestigious university.

The *Russian Area Studies Program* provided a comprehensive interdisciplinary approach to studying the Soviet Union and Russia. Students in the program were encouraged to conduct original research and produce a dissertation on a topic they chose within Russian Area Studies. Faculty members were experts in various disciplines, offering guidance and mentorship to students throughout their academic journey. In January of 1981, I began my doctoral studies at Georgetown. The first course in the curriculum was *"Soviet Military Power."* My professor, Dr. William (Bill) F. Scott, a retired Air Force colonel, was one of the nation's leading authorities on the Russian military. He had served two tours of duty at the American Embassy in Moscow as both Air Attaché and acting Defense Attaché. Having the rare privilege to attend Dr. Scott's informative lectures, I entertained a specific yet confident hope that one day, God willing, I would be sent on assignment to Moscow as the Air Attaché.

During my attendance at Georgetown and the remainder of my Air Force career, Bill Scott and his wife, Harriet, became my mentors, role models, and lifelong friends. While Bill was acknowledged as an accomplished Russian expert in both the military and academia, Harriet was fluent in Russian. She was a senior research associate at the Center for Advanced International Affairs at the University of Miami and a consultant on Soviet military and political affairs to several major research organizations. Without the continued support, inspiration, and good counsel of Bill and Harriet, it is unlikely that I would have been able to accomplish my goal of becoming the first African American Air Force Attaché to Moscow.

Although I was successful in my initial coursework at Georgetown, during the second semester, I regretfully had to make a difficult decision. It is famously said that *"God will never give you more than you can* handle." However, the pressures and challenges of working at such a high-powered and high-profile desk at the Defense Intelligence Agency, combined with frequent temporary overseas assignments, my very intensive doctoral studies, and the unavoidable obligation of providing adequate care for my three children as a single parent, ultimately proved too much for me to handle. In the end, I was compelled to arrive at the painful decision to suspend my doctoral studies at Georgetown so that I could spend more time at home with children who needed their father daily. With the wisdom of hindsight, I am convinced this was the best decision for my family and well-being.

In August of 1981, a dramatic change occurred in the trajectory of my life. For her fifty-eighth birthday, I had invited my mother out to dinner at an upscale restaurant on the waterfront in Washington, DC. After dinner, I took my mother back home and returned to a nightclub adjacent to the restaurant. The kids were with a babysitter, and this was one of those few and far-between moments for me to have fun. As I entered the nightclub, I spotted a group of about eight people at a private party celebrating a friend's success at the Washington, DC, Bar Examination. Quite unceremoniously, I insinuated myself into the private party and asked politely if I might join the celebration. The host graciously agreed to my unannounced intrusion into his party, and I found myself seated at a table next to a well-dressed, attractive Black woman with long, flowing hair. I quickly introduced myself, and we started a conversation. Her name was Annie, and she was an elementary school teacher who had moved to Washington, DC, from South Carolina several years earlier to continue her teaching career.

As it turned out, Annie and I had much in common. We were both divorced and were single parents. Annie had a young son named Torrey, and I had three children named Steven, Marten, and Karen. At the end of the evening, I gave Annie my business card, which contained my name, *Lewis S. Wallace Jr., Captain, US*

Air Force, and telephone number. Annie reciprocated and gave me her telephone number. The next day, I called Annie and invited her for dinner at the Officer's Club at Bolling AFB. She readily accepted my invitation, and we arranged to meet the following Saturday. All week long, I kept thinking about Annie. I knew that I found her very attractive. I also knew that our conversation at the nightclub was most stimulating. Yet, I also realized just how vulnerable I was, having gone so long without nurturing female companionship. However, I felt a curious kind of safety in my thoughts about Annie and decided to plunge right ahead into socializing with her. Little did I know that my chance encounter with Annie in a nightclub was about to alter the course of my life in many more remarkable ways than one.

Before dinner at the Officers' Club, I took Annie to my house to meet my children. After introducing Annie to the kids, something surreal, if not perfectly strange, happened. Steven, Marten, and Karen immediately approached Annie and warmly embraced her. From that point, I knew that God had sent me this beautiful woman to become an integral part of my life and my children's. I had to turn my face away to dab at the tears in my eyes. Over the next several months, Annie would become a constant feature of the landscape of my life, and together with our four children, we all became inseparable. Every weekend, we would take the kids on sightseeing outings to museums, recreation parks, and church services on Sundays. When I finally introduced Annie to my entire family, they became very fond of her, and my mother quietly whispered to me, *"She is a keeper!"* Annie told me she was equally fond of my family and hoped we could visit South Carolina with her parents soon.

At this point, without hesitation or further delay, I professed my eternal love for Annie and asked her to be my wife. She kissed me tenderly and readily accepted my marriage proposal. Annie also told me that she had prayed to God to send her an angel who would love and protect her and her son for the rest of her life, and I was that angel. After our engagement, Annie and I had a serious conversation about the future. I told her about my dream of becoming the first African American air attaché at the American Embassy in Russia. I asked her how she would feel about accompanying me on this incredible journey. To my surprise, Annie responded that although she was a "country girl" from the small town of Union, South Carolina, and had never even been outside of the United States, she loved me dearly and would be honored to accompany me anywhere in the world as long as we could stay together as a family.

After receiving that positive and enthusiastic response from Annie, I immediately contacted the Air Force Officer's Personnel Center to find out the requirements set by the Air Force for the position of Air Force Attaché to Moscow. The requirements included proof of citizenship, rank and experience prerequisites, foreign language proficiency, and educational qualifications. A month after submitting my application, I received a call from the Officer Assignment Branch that my application for attaché duty was unsuccessful. When I asked why I had been denied, the assignments officer responded that, despite my excellent qualifications, they selected a more highly qualified officer for the position. When I asked what I could do to become "more qualified," the response was very explicit. In a word, I could do *nothing*. Dumbfounded by this unfortunate news, I graciously thanked the assignments officer for his call and said I would reapply for the attaché position sometime soon. I knew that through persistence and determination, one day, God willing, I would be successful in my quest to become an attaché.

In October 1981, Annie and I received tickets to attend the annual Air Force Officers' Charity.

Ball at Bolling AFB. That evening, we had the unexpected pleasure of speaking with Air Force Brigadier General Donald W. Goodman, the assistant vice director for attachés and training at the Defense Intelligence Agency. I told General Goodman of my desire to become an Air Force Attaché in Moscow and briefly outlined my qualifications. My outstanding record of achievement as an aircrew member and Russian Area Specialist impressed him. When I introduced General Goodman to Annie, he asked about her profession. Annie proudly responded that she was an elementary school teacher who grew up in a small town in South Carolina. The general smiled and said that Annie would be the perfect partner to accompany me to Moscow if I could get the job. After those encouraging words of inspiration from General Goodman, I knew nothing could thwart my dream of someday becoming an Air Attaché in Moscow.

In February 1983, I received a call from my friendly Air Force Assignments Officer. He said that my three-year tour of duty in Washington, DC, would end in July, and I was reassigned to Headquarters, Tactical Air Command, as Chief of the Warsaw Pact Branch. He also informed me that I might soon be on the promotion list for Major (O-4). Of course, I was thrilled about the prospect of attaining a higher rank. Yet, I was not thrilled about leaving my fiancé, family, and friends and relocating to Hampton, Virginia. When I told Annie about these unfortunate events, she smiled and said, *"God will work everything out."*

With little time to waste, on April 1, 2003, April Fool's Day, I called Annie from my office and told her to put on her white dress and meet me at the courthouse in Fairfax, Virginia, as soon as possible because we were getting married that day. Initially, Annie thought this was an April Fool's joke. However, when Annie realized I was dead serious, she happily agreed to meet me as soon as possible. After completing the required paperwork, we located a Justice of the Peace near the courthouse and exchanged our wedding vows. Before we departed Washington, DC, for my new assignment in Hampton, Virginia, we arranged a wedding reception for our families and friends at the Officers' Club. In July 1983, Annie, our four children, and I moved to Langley AFB, Virginia, ending another chapter in my incredible odyssey.

CHAPTER TWENTY

ON THE THRESHOLD OF ACTUALIZING MY DREAM

"Intelligence without ambition is a bird without wings."
- Salvador Dali

Although I initially had reservations about leaving my high-profile job at the Defense At the Intelligence Agency in Washington, DC, for a new position at Langley Air Force Base in Hampton, Virginia, I ultimately concluded that it was a beneficial move, since one of the prerequisites for sustained and consistent promotion in the Air Force was to serve credibly in a variety of assignments. In my case, I could showcase the analytical, leadership, and management skills I gained from these diverse career experiences to the many different specialized units of the United States Air Force. Additionally, having previously served at Headquarters Strategic Air Command at Offutt AFB, Nebraska, I knew that being assigned to Langley AFB, Virginia, would provide me an excellent opportunity to deepen my understanding of the tactical employment of air power to support ground forces in various combat scenarios.

In early July of 1983, Annie, our four children—Steven, Marten, Karen, and Torrey—and I made the 200-mile journey from Washington, DC, to Hampton, Virginia, to begin our new life as a blended family. Hampton, Virginia, is a vibrant and growing city in the southeastern part of the state along Chesapeake Bay. The city has a rich history, dating back to the early 1600s, and is known for its military presence, as it is home to Langley AFB, NASA Langley Research Center, and the Virginia Air and Space Center. Featuring many miles of enchanting waterfront and beaches, Hampton also hosts numerous businesses and industrial enterprises, as well as retail and residential areas and sites of historical interest.

After completing my in-processing at the Headquarters Tactical Air Command (TAC) Administrative Office, we checked into temporary family quarters on base until I found a more permanent housing solution for my family in the Hampton area. After contacting about four realtors in the local area, we finally settled on and purchased a four-bedroom home in the Willow Oaks subdivision, approximately ten miles from the base. Having settled my family in our new residence, my next priority was reporting to my new duty station. When I arrived at the Headquarters TAC Intelligence Division facility, I was warmly greeted by several officers and enlisted personnel in the unit. They complimented me on my previous outstanding performance as both an Intelligence Analyst and Russian Area Specialist and wished me well in my new assignment.

To my utter surprise, however, the secretary informed me that the Assistant Deputy Director for Operational Intelligence, an Air Force colonel, sought an immediate audience with me in his office. Upon entering his office, I saluted him smartly and took a seat. The colonel said he was pleased to have an officer with such strong qualifications on his staff. He informed me that I was assigned as the Chief of the Warsaw Pact Branch, a position generally held by a Major. I was initially perplexed by the colonel's announcement. I immediately thought, *"Why would a captain be assigned to fill a position reserved for a major?"* Almost as if he could read my mind, the colonel smiled and said he was fully confident I could handle the increased responsibility. Finally, the colonel told me that he had received notification from the Air Force Personnel Center that my sequence number for promotion to Major, O-4, had come up, and I would be immediately promoted to Major in the United States Air Force. Overjoyed, I thanked the colonel for sharing the good news, smartly saluted, and departed his office. When I returned home after my first day on the job, I told Annie and the kids about my new position and promotion to Major. They were ecstatic. I scheduled the promotion ceremony for the following week, and Annie proudly pinned Major's gold leaf insignia on my shoulders while the kids loudly applauded their dad's achievement.

As Chief of the Warsaw Pact Branch, my primary duty was supervising a team of intelligence specialists who provided assessments and briefings to the Commander of the Tactical Air Command and its senior staff on military and political developments in the seven Warsaw Pact countries: the Soviet Union, East Germany, Poland, Czechoslovakia, Hungary, Romania, and Bulgaria. The Warsaw Pact, formally known as the Treaty of Friendship, Cooperation and Mutual Assistance

(TFCMA), was a collective defense treaty signed in Warsaw, Poland, between the Soviet Union and seven other Eastern Bloc socialist republics of Central and Eastern Europe in May 1955, during the Cold War. The term "Warsaw Pact" commonly refers to the treaty and its resulting military alliance, the Warsaw Treaty Organization (WTO). Dominated by the Soviet Union, the Warsaw Pact was established as a balance of power or counterweight to the North Atlantic Treaty Organization (NATO) and the Western Bloc. There was no direct military confrontation between the two organizations; instead, the conflict was fought on an ideological basis and through proxy wars.

At the end of 1983, while still working at Headquarters Tactical Air Command, my career took another turn. I applied and was selected by the Dean of the College of Arts and Sciences at Hampton University as an adjunct professor to teach an evening basic Russian course to ten engineers working on the US Space Shuttle Program at NASA's Langley Research Center. This was a groundbreaking course for Hampton University, a Historically Black College and University (HBCU), because the Foreign Language Department at Hampton had historically only offered courses in the traditional foreign languages of French, Spanish, and German. The basic Russian course I taught achieved remarkable success; after only two semesters of study, all of my students made significant progress in reading and translating various Russian articles. Due to my success in teaching basic Russian courses, the Dean of the Faculty of the College of Arts and Sciences at Hampton University asked me if I would consider leaving the Air Force to accept a position as a professor of Russian at Hampton University. Although I was intrigued by the thought of teaching full-time at the university level, I politely declined the dean's gracious offer, telling him that I wished to continue my career as an Air Force officer, one that, with each passing day, gleamed with the prospects of a bright future. I also informed the dean of my dream of becoming the first African American assigned to the American Embassy in Moscow as an Air Force Attaché.

After a long conversation with Annie about my future in the Air Force, I resubmitted my application to the Air Force Attaché Office for the position of Assistant Air Attaché in Moscow.

However, about a month after submitting my application, I received a call from the Attaché Office stating that the Air Force had again denied my application for the assignment. In utter disbelief, I asked the assignments officer, *"Why was my application rejected for the second time, and what can I do to become more competitive for this job?"* He responded, *"Major Wallace, you have an outstanding record of achievement in some of the most critical jobs in the intelligence field. However, we selected another officer with better qualifications, and there is nothing you can do to make yourself more competitive for this position."* Although I found the explanation unsatisfactory, I swallowed my disgust and thanked the assignments officer for his candid response. I informed him that I would reapply for the position in the future. I thought, *"Maybe the third time would be the charm."*

In July 1984, my Air Force career at Langley AFB took an unexpected turn. One day, the TAC The Director of Intelligence called me into his office. He informed me that I was being reassigned to the Headquarters TAC Inspector General's Office as the new Chief of the Intelligence Inspection Branch. He also noted that in this new position, I would ensure that intelligence functions within the Tactical Air Command met operational requirements and adhered to established standards and best practices. Although I was thrilled at the prospect of enhancing my intelligence portfolio by conducting operational readiness inspections and management effectiveness inspections of intelligence activities and programs within TAC units, the immediate drawback of this new assignment was that I would be away from my family for extended periods while traveling with the inspector general's team. Yet, a significant bonus of my new position was that I was once again placed on flying status and awarded my officer aircrew wings. When I told Annie about my new job, she was incredibly supportive and said, *"Don't worry about me and the kids. God will work everything out for us."* Her response was a perfect example of a supportive wife, and I couldn't help but express my gratitude to her for her reassuring and encouraging attitude. I told myself I was indeed a lucky man.

In June 1985, after serving a year as the Chief of the TAC Intelligence Inspection Branch, I contacted my friendly Air Force Assignments Officer to apply for a Russian Area Specialist position with the Headquarters Air Force Intelligence Agency at the Pentagon. To enhance my qualifications and improve my chances, I needed to return to Washington, DC, as soon as possible for the Assistant Air Force Attaché position in Moscow. Initially, the assignments officer was reluctant to approve my transfer back to the Pentagon. However, after receiving several recommendations from Air Force General Officers supporting my request for a transfer back to Washington, DC, the assignments officer finally relented and offered me a position as the Chief of the Space and Missile Branch in the Weapons, Space, and Technology Division within the Air Force Intelligence Service.

Once again, my perseverance and determination were instrumental in obtaining my new assignment. In August 1985, my family and I packed our belongings and moved to Springfield, Virginia, a suburb of Washington. Our friends and family were delighted that we were once again living near them. Annie accepted a position as an elementary school teacher at a prestigious private school in Virginia, and we enrolled the children in highly rated Virginia middle and high schools. In my new position at Headquarters Air Force Intelligence Service (AFIS), I had the honor of briefing the United States Air Force Chief of Staff and his senior staff on several occasions. I received laudatory comments on my expertise as a Russian specialist. In my 1986 performance evaluation report, my supervisor wrote the following *statement: "Major Wallace has done impressive work in defining the threat to Air Force systems such as Peacekeeper, Small ICBM, and the B-1B (long-range strategic bomber aircraft) from Soviet ballistic missile systems and space-based threats."*

During my tenure as the Chief of the Space and Missile Branch at Headquarters, I had the opportunity to return to Moscow on a temporary duty assignment. While at a reception for high-ranking Soviet officials at the US Ambassador's residence, I introduced myself to the current Air Force Assistant Air Attaché, an Air Force Major. During our conversation, a Soviet General Officer approached us and asked the attaché several questions in Russian. To my surprise, the attaché, who had been selected for his position instead of me, ostensibly because of his outstanding and supposedly superior qualifications, could not understand what the Soviet general was saying. The attaché quickly turned to me and asked if I could interpret their conversation. I immediately introduced myself to the general in Russian and began interpreting for the two. After their discussion, the Air Force Major thanked me profusely for helping him communicate effectively with the Soviet general. However, I was nonplussed that this Air Force Assistant Air Attaché was assigned to Moscow without even rudimentary fluency in the Russian language. Upon my return to Washington, DC, I again applied for the Assistant

Air Attaché position in Moscow. This time, however, I reached out to the network of general officers and senior executive service intelligence professionals I had worked with in the past to personally seek their support and endorsement of my application, sincerely and fervently entertaining the hope that *"the third time would be the charm."*

CHAPTER TWENTY-ONE

THE THIRD TIME WAS THE CHARM

"The greatest glory in living lies not in never falling, but in rising every time we fall."
- Nelson Mandela

The third time was the charm, an idiomatic expression meaning that success or achievement occurred after two previous failures or attempts. Originating from ancient folklore and fairy tales, this phrase emphasizes perseverance and determination. The earliest traces of the expression are in 17th-century English folklore, where the phrase's earliest recorded use was in the 1678 book, "The Nursery Rhymes of England" by James Orchard Halliwell. The phrase's evolution was likely influenced by the appearance of the *"rule of three"* in ancient mythology, such as the *three wishes* granted by genies or gods, and in Christianity, where the *Trinity* and the three-day resurrection narrative might have shaped the concept. Whatever its origins, by the 20th century, it had become popularized in American culture through literature, film, and media. Of greater personal relevance to my incredible life journey, "third time was the charm" would become my reality in 1986.

After two previous attempts to secure the position of Assistant Air Attaché in Moscow in 1986, I decided to submit my application for the third time. I found it difficult to understand why a candidate with my outstanding academic qualifications and impressive career pedigree in the United States Air Force would persistently receive rejection notices from the Air Force Attaché Office. Throughout my military career, I firmly believed that the Air Force had a strong commitment to equal opportunity and diversity in assignments, promotions, and career development. However, due to the roadblocks I encountered in obtaining the Assistant Air Attaché position, I had begun to entertain serious doubts about the Air Force's commitment to diversity, equal opportunity, and inclusion in officer assignments.

Finally, lacking any hard evidence, I arrived at the rather convenient yet entirely plausible conclusion that the primary reason for my non-selection for the attaché position was my race as someone of African American descent. Although numerous African American military attachés had served in various embassies worldwide, historically, the United States had never approved an African American officer to serve as a military attaché to the Soviet Union. As I mentioned in the previous chapter, my renewed approach was a carefully orchestrated game plan designed to enhance my chances of winning the nomination for the attaché position in Moscow. To that end, I decided to "pull out all stops" and solicit the support of several high-ranking general officers and civilian senior executive service intelligence professionals I had previously worked with. These individuals could attest to my qualifications to successfully serve as an Assistant Air Attaché in

Moscow, regardless of my skin color.

As fate would have it, the primary supporter and advocate of my application to become an Air Force attaché was Major General Charles R. Hamm. General Hamm, a former member of the Air Force air demonstration squadron, the *Thunderbirds*, and as a colonel, was the Commander of the Eighth Tactical Fighter Wing at Kunsan Airbase, South Korea, during my tour of duty as the Chief of the Air Targets Branch at Kunsan in 1975. On several occasions, the colonel mentored his junior officers on potential career opportunities in the Air Force. I briefly highlighted my previous assignments as a Russian Area Specialist during one such session. I told him of my desire to someday serve as an Assistant Air Attaché at the American Embassy in Moscow. General Hamm responded, *"Lieutenant Wallace, if you continue to improve your leadership skills, Russian language proficiency, and experience in air operations and international relations, I believe you will be an excellent candidate for the position of Assistant Air Attaché in Moscow."* I thanked the colonel for his kind words and assured him I would work hard during my career to improve my chances of obtaining my dream assignment.

Ten years later, in 1985, I contacted General Hamm, who served as the Deputy Director of Plans at Headquarters U.S. Air Force at the Pentagon, before resubmitting my attaché application for the third time. When I contacted the general's office to request a meeting, his executive officer, Colonel Williams Stevens, an African American, was initially reluctant to approve my request. He suggested that I first meet him in his office to disclose the nature of my business with his boss. I briefly told Colonel Stevens that I had worked with General Hamm in Korea and requested his support and endorsement of my application to become an Air Force Attaché in Moscow. During our meeting, Colonel

Stevens, to my surprise, told me that he had previously served as air attaché to Liberia, Sierra Leone, Ghana, Niger, Upper Volta, and the Ivory Coast. He also told me that he fully supported my application to become the first African American air attaché in the Soviet Union, and he would work diligently to convince his boss to endorse my application. During my subsequent meeting with General Hamm, the General told me he would write a letter to the Air Force Attaché Office endorsing my application for the position in Moscow. Undoubtedly, the general's endorsement would carry enormous weight with the Attaché selection panel because he had previously served as the U.S. Defense Attaché to the Soviet Union from 1981 to 1983.

Sometime in 1986, I formally submitted my application for the position of Assistant Air Attaché in Moscow. I truly believed that the *"third time would be the charm,"* and because of my determination, persistence, and hard work, I would finally be selected to attend an interview for this position. After formally submitting my application, I met with my boss, an Air Force Brigadier General serving as the Deputy Director for Air Force Intelligence at the Pentagon. He wished me well in my quest and said, *"Major Wallace, I successfully got you an interview with the attaché selection panel for a possible attaché assignment to Moscow. However, using baseball vernacular, I must let you know that I got you a bat, and now it is up to you to hit a home run."* I thanked the general for his words of support and encouragement and responded that my wife, Annie, and I would indeed hit a "home run" during our interview.

In June 1986, I received a call from the attaché office that I had been selected for an interview for the position of Assistant Air Attaché to Moscow. Annie and I were ecstatic about this good news. In preparation for the interview, I thoroughly researched current developments in Soviet military, political, economic, and cultural affairs. Additionally, both Annie and I met on several occasions with Russia experts Colonel Bill Scott and his wife, Harriett, who had served two previous tours in Moscow as the Air Attaché and Acting Defense Attaché, to learn first-hand about some of the unique challenges confronting an attaché family while stationed in Moscow. Before our interview, Annie and I went shopping at upscale department stores in the Washington, D.C., area to purchase new clothing to wear to the interview. To be selected for this assignment, we both had to "dress for success" and convince the selection panel that we would be the perfect attaché-wife team to represent the United States in the fast-paced diplomatic environment at the American embassy in Moscow.

After the two-hour interview, the selection panel chairman congratulated Annie and me for our outstanding preparation. He remarked that we were the most qualified candidates he had ever interviewed for this position. In addition, the chairman said that the panel would unanimously forward my nomination to the requisite Department of Defense offices for acceptance of my assignment to Moscow. Of course, Annie and I were overjoyed at this exceedingly good news. Finally, after fourteen years as an Air Force officer, I had attained my goal of being selected for assignment as an Assistant Air Attaché in Moscow. In hindsight, I recalled the prophetic words of my grandmother and mother, through whom I had become familiar with the biblical *verse, "No weapon formed against Thee shall prosper."* That enduring scripture had become a statement of my reality.

My sojourn in the United States Air Force had been a study of perseverance. Indeed, perseverance is a quality that has defined my journey through life. Growing up, I faced numerous challenges that tested my resolve, but each obstacle strengthened my determination to succeed. Despite initial difficulties in many different facets of my life, I persevered, seeking help and utilizing extra resources whenever necessary. Professionally, perseverance has been crucial in my career trajectory. Rejections and setbacks were inevitable, but I chose to learn from them, with every "no" bringing me closer to the *"yes"* I sought. In sober reflection, perseverance has been the thread that weaved all my experiences together to form the wonderful fabric of accomplishment that has been my life story. Perseverance has taught me resilience, adaptability, and the value of persistence. I've come to understand that success is not solely defined by achievements but by the courage to continue in adversity. As Nelson Mandela, the 1993 recipient of the Nobel Peace Prize for his efforts to dismantle his country's apartheid system and South Africa's first Black president, once said, *"The greatest glory in living lies not in never falling, but in rising every time we fall."* My life's journey has embodied this spirit, transforming obstacles into steppingstones for growth. My incredible journey continues.

CHAPTER TWENTY-TWO

ATTACHÉ TRAINING AND ORIENTATION

"Effective performance is preceded by painstaking preparation."
- Brian Tracy

As soon as I received Headquarters Air Force approval of my nomination to become the Assistant Air Attaché in Moscow, I began specialized training to prepare for the assignment. An essential part of my training was attendance at the Joint Military Attaché School (JMAS). This course was designed to prepare military officers for the role of a military attaché. The curriculum included a combination of academic lectures, practical exercises, and simulations to provide participants with a comprehensive understanding of the duties and responsibilities of a military attaché. On December 12, 1964, the Defense Attaché System (DAS) was established as an arm of the Defense Intelligence Agency and tasked with representing the United States in defense and military-related matters with foreign governments worldwide. Defense Attaché Offices (DAO) operate from US embassies globally. DAOs are composed of civilian and military employees, most of whom receive specialized training at the Joint Military Attaché School before their appointment.

Membership in the Defense Attaché Service (DAS) means assignment to one of the most elite duties in the Department of Defense (DoD), and it begins with training at the Joint Military Attaché School (JMAS) for the many military and civilian attachés and attaché support personnel for a critical international duty. Serving at a US embassy comes with significant privileges and responsibilities, and attachés and their families are among the select few who represent the capabilities and traditions of the US Department of Defense abroad. The JMAS training program was demanding and fast-paced, involving constant evaluation throughout the course. We were challenged in realistic and, at times, stressful training situations. As highly experienced military personnel, the training program took into account our experience and knowledge base and leveraged it with a specific training curriculum. The training program utilized a team-based instructional approach and practical field exercises to prepare us for the upcoming DAS assignments.

While the JMAS provided the tools and knowledge needed for a successful Defense Attaché Service (DAS) tour, succeeding at JMAS required one to employ the honor, dedication, and attention to detail that had brought one to this point in one's career. Also, because an attaché assignment is a "family affair," Annie had to attend specific seminars and workshops at the Foreign Service Institute and the Joint Military Attaché School to help her navigate the challenges of diplomatic life and represent the United States effectively while living in the Soviet Union. A few courses that Annie attended included cultural awareness and sensitivity training, protocol and etiquette training, introductory Russian language classes, security briefings, and familiarization with Russia's customs and traditions.

Two of our children, Karen and Torrey, both teenagers, initially had reservations about leaving their friends, family, and schools in the United States to move to faraway Moscow for two years. However, I educated them on some of the positive aspects of living abroad and how being part of the diplomatic community would be an opportunity of a lifetime for two African American youngsters. Finally, they relented and became somewhat enthusiastic about moving to Russia. As part of their orientation before departing for Moscow, Karen and Torrey attended the State Department's *Young Ambassadors Program* for children of US diplomats. The program provided them with a unique opportunity to gain a better understanding of diplomacy and international relations. More importantly, the Young Ambassadors Program allowed Karen and Torrey to connect and interact with children from other diplomatic families and learn about embassy life in different countries.

In the past, many US military attachés were typically of a higher rank than I. They had attained a certain level of wealth and material comfort during their military careers. However, it was a far more challenging scenario for a major with a wife and four children, two of whom were in college. Air Force and Department of Defense regulations stipulated that all attaché designates must be financially stable before taking up their diplomatic assignments. In my case, obtaining the necessary funds for a tour of duty to Moscow was no easy task. The estimated start-up cost for my family's move to Russia was approximately twenty-five thousand dollars in 1988 currency terms. That amount included the purchase of service uniforms, aiguillettes, mess dress uniforms, business suits, cocktail dresses, and gowns suitable for official and social functions. Additionally, because Moscow regularly experienced frigid winters with heavy snowfalls, I had to purchase a substantial amount of cold-weather clothing for the entire family. Since Western products were generally unavailable in the Soviet Union, we could ship up to 3,000 pounds of consumable goods to Moscow in addition to our household goods allowance. We had to estimate how many rolls of toilet paper, towels, washing detergent, cleaning supplies, toothpaste, shampoo, spaghetti, sugar, soft drinks, canned goods, and various other items we would require for our two-year tour of duty in Moscow. To accomplish this Herculean task, Annie and I had to shop at the local base commissary on a Sunday afternoon when the store was closed to the general public. That was the only way we could conveniently purchase our goods in bulk for onward transport to Moscow.

Success as an Assistant Air Attaché necessarily demanded far more than merely passing, if not excelling, in the Russian language. In any case, the Defense Attaché Service (DAS) of the Department of Defense was insistent that all attachés be able to communicate effectively in the language of the host country where they would be serving. It was purely a matter of diplomatic expediency, and in my case, the crucial role of Russian language proficiency was not negotiable. The Assistant Air Attaché is a critical diplomatic role requiring exceptional skills, particularly in language proficiency. To effectively represent US interests and facilitate international cooperation, excellence in Russian language skills is essential. Language mastery meant fluency in reading, writing, speaking, and understanding Russian.

Additionally, cultural competence demands in-depth knowledge of Russian customs, protocols, and nuances. Naturally, my stint at the Joint Military Attaché School (JMAS) had given me the diplomatic acumen that all but guaranteed my ability to navigate complex international relationships. My technical expertise was already established as a major in the USAF through my familiarity with aviation and military affairs. Significantly, as an Assistant Air Attaché, I would be required to engage in effective negotiation and representation, deploy my linguistic skills for accurate translation and interpretation, and be a strategic analyst capable of interpreting Russian military doctrine and strategy.

When all was said and done, to succeed as an Assistant Air Attaché, exceptional Russian language proficiency was indispensable to me. Mastery of the language, cultural competence, and diplomatic acumen would be essential for effective representation, negotiation, and cooperation. I was very pragmatic in my approach to my imminent assignment. Although I had spent many years learning Russian, to enhance my effectiveness as a Russian linguist while assigned to Moscow, I requested an assignment to the one-year advanced Russian language program conducted by the US Army Russian Institute in Garmisch, Germany, before departing for Moscow. To my utter amazement, the Air Force granted my request, and in June of 1987, Annie, the children, and I departed Washington, DC, for Garmisch, Germany, to inaugurate a whole new phase of my saga.

CHAPTER TWENTY-THREE

THE SKI SLOPES OF GARMISCH

"It doesn't matter where you are coming from. All that matters is where you are going."
- Brian Tracy

As we boarded our plane at Washington Dulles Airport for the eight-hour flight to Frankfurt, Germany, Annie and the kids were filled with both anxiety and excitement. Being the first time my wife and children would be traveling overseas, they asked me numerous questions throughout the flight about what it would be like to spend a year living in Germany before going to Russia. Because I was an active-duty officer flying on official military travel orders, we had the good fortune of being upgraded by the airline to "Business Class" for the flight to Frankfurt. At Frankfurt Airport, we made our connections for our flight to Munich. Because Munich was located approximately 85 miles from our final destination, I rented a car at the airport for the last leg of our trip to Garmisch.

Garmisch is a picturesque resort town in Bavaria, Germany, known for its stunning alpine scenery. It is a popular destination for outdoor enthusiasts, offering activities such as skiing, snowboarding, hiking, and mountain biking in the nearby Alps. It was the site of the 1936 Winter Olympics, the first to feature alpine skiing. The city later replaced Sapporo, Japan, as the host of the 1940 Winter Olympics, which, however, was ultimately canceled due to World War II. The town is also home to the famous *Zugspitze,* Germany's highest peak, which offers breathtaking views and winter sports opportunities.

Upon arriving in Garmisch, I reported for duty at the US Army Russian Institute to begin a one-year *"special student"* program of instruction in advanced Russian in preparation for my assignment to Moscow as the Assistant Air Attaché. Notably, I had the distinct honor of being the first Air Force officer and the only person of minority background to be enrolled as a full-time student at the US Army Russian Institute. The institute conducted two years of advanced university-level training for the Army's Foreign Area Officer Program for the USSR and Eastern Europe. The program included required and elective courses in Russian/Soviet geography, history, foreign policy, government, economics, literature, military doctrine and force structure, and language. Virtually all course offerings were taught in Russian by native faculty.

To supplement formal classroom instruction, students were taken on orientation trips to Eastern Europe and, when possible, the Soviet Union. Before leaving Washington, DC, Annie applied for and was hired as an elementary school teacher at the Department of Defense Dependent Schools (DoDDS) in Garmisch. The school served the children of US military personnel and government employees stationed there. It offered a curriculum that met US academic standards and emphasized college and career readiness. Additionally, the school provided a supportive learning environment for kindergarten through twelfth-grade students, focusing on academic excellence, character development, and preparation for future success. Karen and Torrey made friends with their classmates and quickly adapted to their new school environment. However, they sometimes found it awkward attending a school where their mom was a teacher.

In addition to core academic subjects, the Garmisch school offered special extracurricular activities that reflected the region's unique cultural features and historical context. My children had the opportunity to participate in sports clubs and other activities that contributed to their overall development and enriched their educational experience. On a practical note, classes were suspended every Wednesday during winter, and students were given complimentary lift passes to go skiing. Although Annie was not a fan of outdoor winter sports, Karen and Torrey became proficient in skiing and ice skating during our time in Garmisch.

Although I excelled in my Russian classes, being the only Air Force officer-student in a class of forty Army officers presented a challenge. For example, the US Army adopted the Battle Dress Uniform (BDU) to replace the previous olive-green fatigue uniform as the uniform of the day for all Army students attending the institute. The Battle Dress Uniforms (BDU) have undergone significant transformations since their introduction in the mid-20th century. Initially designed solely for Army infantry personnel, BDUs evolved to meet changing combat requirements, technological advancements, and practical needs. In the 1960s-1970s, the US Army had the iconic OG-107 Cotton Sateen Uniform, which later gave way to the BDU featuring a woodland camouflage pattern. My major dress constraint during my tour of duty at Garmisch was that my service branch, the US Air Force, had not yet transitioned to a new combat uniform. I arrived at the station in my old, olive-green fatigue uniform, which led my classmates to often bully me for being out of uniform and not adhering to official Army uniform standards.

On one occasion, my class leader, a US Army Major, reported me to the installation commander, an Army Colonel, for insubordination because I had refused to purchase the US Army Battle Dress Uniform. Naturally, I was summoned to meet with the commander, where I explained that although I was honored to be the only Air Force student at the institute, I was not bound by the US Army's dress code and considered the constant harassment by the class leader and other students unacceptable behavior. Additionally, I informed the commander that I had just received a phone call from the Air Force Personnel Center stating that I was on the selection list for promotion to the rank of Lieutenant Colonel (O-5). I concluded our meeting by requesting to be appointed the senior class leader in light of my upcoming promotion. Although the installation commander promptly rejected my proposal, I was no longer bullied by the class leader and the other students for being *"out of uniform."*

After graduating from the advanced Russian language course at Garmisch in June of 1988, I planned to continue our journey eastward to Moscow without first returning to Washington. However, Annie and the kids were adamantly opposed to my proposal. They informed me, in no uncertain terms, of their desire to return to Washington for a final farewell with family and friends before departing for Moscow. Faced with this strong opposition from my family, I finally agreed to their demands, and we flew back to Washington, DC, for a three-week vacation. During this welcome interlude from service work and training,

I received the exciting news that President Ronald Reagan had approved my assignment as the Assistant Air Attaché at the American Embassy in Moscow. After completing the final rounds of consultations and briefings with various government agencies and receiving our diplomatic passports, we departed Dulles Airport in early July for our trip to Moscow.

With the president's approval of my appointment as an Assistant Air Attaché, I finally achieved the service ambition I had harbored for many years. For me, achieving my ambition not only perfectly validated my mother's words, *"Son, in this great country, you can be anything you want to be. It just takes believing in yourself, getting a good education, working hard, and never giving up on your dreams,"* but also brought a mixture of immense pride, a sense of accomplishment, relief, and even a little disbelief. Quite frankly, it felt as if I had finally arrived at a long-awaited destination after a challenging journey. It was a profoundly fulfilling and validating experience. However, it also left me reflecting on what I would aspire to in my life trajectory as soon as this initial euphoria subsided. Ultimately, I found contentment in a strong feeling of satisfaction and self-worth for having accomplished this significant goal. I also felt the release of tension and stress from pursuing such a challenging ambition while being secure in the belief that hard work and dedication will always eventually pay off. As Annie, the kids, and I journeyed to Moscow, what remained paramount in my consciousness was a strengthened belief in my ability to tackle future challenges, so help me God.

CHAPTER TWENTY-FOUR

CULTURE SHOCK AT MOSCOW AIRPORT

"Here are luxury and penury, abundance and the most extreme deprivation, piety and atheism, and an unbelievable frivolity; warring elements which, out of their constant conflicts, create this marvelous, outrageous, gigantic whole which we know by its collective name: Moscow."

- **Konstantin Batyushkov,** in 1982, as quoted in City Journal (2013)

On July 15, 1988, Annie, the children, and I departed Washington Dulles Airport on a Pan American Airways airliner for our flight to Moscow. Because we traveled on diplomatic passports and had entry visas into the Soviet Union, the airline gave us courtesy upgrades to *Business Class*. Our flight itinerary would take us to Moscow, with an intermediate stopover at Frankfurt Airport in Germany. During the nine-hour flight from Washington to Frankfurt, Annie, Karen, and Torrey were noticeably quiet and seemed to be deeply contemplating what life would be like in the distant land called the Soviet Union.

We arrived at Frankfurt Airport and proceeded to the ticket counter to check in for our three-hour flight to Moscow. As he reviewed our credentials, the airline representative could barely conceal his puzzled look upon discovering that this family of four African Americans had tickets to fly into Moscow. He asked us directly if we were sure we wanted to fly to Moscow. I assured him that we were traveling with diplomatic status and were indeed flying to Moscow. The airline representative had every reason to be astonished. Traveling to the Soviet Union during that era was not a popular adventure among the general population from the West. The reasons for this reluctance to engage in Soviet tourism included the restrictive political climate, limited personal freedoms, potential for surveillance, difficulty in accessing information, shortages of consumer goods, and the risk of being caught up in political dissent or even arbitrary arrests, all stemming from the communist regime's tight control over society.

Even then, stories were rife of tourists being restricted to designated areas and heavily monitored by authorities, with limited ability to explore freely or interact with locals outside official channels. Additionally, access to information about the country's political situation or criticism of the government was heavily controlled, making it difficult to obtain a complete picture of life in the Soviet Union. Such heavy censorship all but guaranteed a strong potential for harassment or arrest, as expressing political dissent, even casually, could lead to interrogation or detention by the authorities. To compound matters for the average visitor to the Soviet Union, there was an acute lack of consumer goods, with severe shortages of essential items like food and clothing. Ultimately, the general lack of transparency complicated the true nature of daily life and social issues due to the government's propaganda and control over information. Bureaucratic hurdles of any kind worsened this tendency.

Having put the airline representative at ease about our more reassuring travel status, we boarded the flight to Moscow. The flight time from ·Frankfurt to Moscow was approximately three hours. Upon landing at Moscow's Sheremetyevo International Airport, I was struck by a familiar sight. On the tarmac, our aircraft was surrounded by military armored vehicles and KGB border guards. Also, as we exited the plane, uniformed KGB guards bearing ominous-looking submachine guns were stationed on both sides of the hallway leading to Passport Control. Because of my previous experience with East German security officers during my first trip to Moscow, I was not alarmed by the high-level security at Sheremetyevo Airport. Not surprisingly, however, Annie and the kids were visibly traumatized and shaken as we approached Passport Control. They had never encountered this level of intimidating security at any airport in the United States.

At Passport Control, the KGB officer took thirty minutes to painstakingly review our passports and visas before finally stamping them and allowing us to enter the Soviet Union. In the arrival hall, two Defense Attaché Office staff members warmly welcomed us to Moscow. They took charge of our luggage and drove us to the American Embassy. Because our permanent quarters were not ready for occupancy, we were assigned temporary housing in the embassy complex. After settling into our temporary lodging, I talked with Annie and the kids about their first impressions of Moscow. Their response was not entirely unexpected. Although they had been raised in the United States and had spent a year in Germany, nothing could have prepared them for the fear and anxiety they felt upon entering the Soviet

Union for the first time. I did my best to allay their concerns by telling them that things would improve during our tour and that they might even take back some fond and wonderful lifelong memories of their time in Russia.

On my first working day at the Defense Attaché Office, I met my fellow attachés from the Air Force, Army, and Navy. Next, I had a thirty-minute meeting with the Defense Attaché, an Air Force Brigadier General. He complimented me on the outstanding credentials that qualified me for the job in Moscow and was pleased to learn that I possessed a high degree of proficiency in the Russian language. He further explained that he had only completed a basic Russian course before coming to Moscow and that he might require my linguistic support when attending some high-level meetings within the Soviet Ministry of Defense. Before departing the United States for Moscow as a new Air Force Attaché, I had been diplomatically accredited by the Soviet Ministry of Foreign Affairs and the Ministry of Defense. However, upon my arrival in Moscow, my first official act was to accompany the Air Attaché to a meeting at the Soviet Military's Foreign Liaison Office to obtain the necessary diplomatic status and privileges. Although the foreign liaison officer was extraordinarily cordial and diplomatic during our one-hour meeting, I could tell that he was somewhat perplexed and taken aback by the ease with which he found himself conversing in both Russian and English with me, the new US Assistant Air Attaché, an African American.

Two weeks after arriving in Moscow, we received a formal invitation to attend the National Day Reception hosted by the Chinese Defense Attaché. Both Annie and I were thrilled that we would finally dress in our civilian attire and put our attaché training to use at an actual diplomatic event. When we arrived at the Chinese Embassy, the reception hall was filled with military attachés from around the world, accompanied by their spouses. When Annie and I approached the Chinese Defense Attaché and his wife for an introduction, I heard them whisper to each other, *"Look at those dressed-up Africans; they speak perfect English."* Ignoring the comment, I congratulated our hosts in English on their National Day and introduced myself as the new US Assistant Air Attaché. After conversing for a few minutes in English, I switched to Russian to continue addressing our hosts. Finally, I switched from Russian to Chinese to thank them for their gracious invitation. Annie then greeted the Defense Attaché and his wife in Russian. The look of utter amazement on their faces was beyond description. It was clear that meeting an African American US attaché couple like Annie and me was a unique experience for them. It was little wonder that after this first meeting, I was able to develop a very cordial and professional relationship with the Chinese Defense Attaché during the remainder of our tour in Moscow.

After spending three weeks in temporary embassy quarters, we finally moved into our permanent residence within the embassy grounds. On-base embassy compound quarters for attachés with dependents were remarkably well-maintained and resembled the two-story townhouses found in the Georgetown area of Washington, DC. To our surprise, the quarters were nicely furnished, well-equipped, and spacious, comprising four bedrooms, two bathrooms, a living area, a kitchen, and other necessary facilities for convenient daily living. Overall, our accommodation in the embassy compound provided a secure and comfortable living environment, offering a sense of security and wholesome community support within the diplomatic community.

CHAPTER TWENTY-FIVE

A SECURITY BREACH AT EMBASSY MOSCOW

"Lonetree's motivation was not treason or greed, but rather the lovesick response of a naive, young, immature, and lonely troop in a lonely and hostile environment."
- **General Alfred M. Gray,** Commandant of the US Marine Corps

The presence of Soviet security guards around the US embassy in Moscow was overwhelming, if not somewhat intimidating. Overall, it reflected the sensitive nature of diplomatic relations between the United States and the Soviet Union. Security measures around the embassy complex included a combination of Russian militia, commonly referred to by embassy residents as *"mili-men,"* security personnel, and surveillance equipment supposedly installed to ensure the safety and security of the entire diplomatic mission. From a Russian official standpoint, these measures were in place to protect embassy staff, visitors, and the premises from potential threats or unauthorized access.

The perimeter of the embassy complex was protected by ten guard stations operated by the militia. However, a contingent of US Marine security guards handled the embassy's interior.

In 1987, the year before we arrived in Moscow, there had been a high-profile espionage case at the embassy involving Marine security guards. The case centered on Marine Corps guard Clayton J. Lonetree, who was arrested and charged with espionage for passing classified information to the Soviet KGB. Lonetree's actions led to one of the most serious security breaches in U.S. diplomatic history. The case highlighted critical vulnerabilities in the screening and supervision of Marine security guards and the risk of espionage in sensitive diplomatic posts.

Indeed, the entire scandal was as rich in its salacious nature as it was revealing of the soft underbelly of America's security apparatus in sensitive foreign missions. Marine Corps guard Clayton J. Lonetree was stationed at the American Embassy in Moscow as part of what was considered a tight security cover for the diplomatic mission in Russia. He was seduced by a twenty-five-year-old Russian woman, *Violetta Seina,* at the annual Marine Corps Ball in November of 1985. At about 5 feet 9 inches tall, with shoulder-length brown hair and striking gray eyes, Violetta Seina always stood out at embassy social functions in fashionable attire. Although employed as a telephone operator and translator for *the Embassy in* Moscow, she lived a double life as a KGB agent. Lonetree was so highly regarded that he was chosen to be part of the Marine unit assigned to provide security for the 1985 summit between Soviet Premier Mikhail Gorbachev and President Ronald Reagan. However, despite the strict *non-fraternization* policy imposed on all Marine security guards (MSGs) in such critical parts of the world, Lonetree and Seina initiated a relationship soon after they met. She introduced him to her *"Uncle Sasha,"* KGB operative Aleksey Yefimov, who asked Lonetree to become a *"friend of the Soviet Union."*

Soon enough, Lonetree was convinced to turn over confidential information, including embassy floor plans. It was alleged that he also gave the names of CIA personnel as well as the identities of U.S. undercover intelligence agents in the Soviet Union, detailing the work habits of embassy staff in the process. After being transferred to the Embassy in Vienna in 1986, he passed on blueprints of that embassy and burned bags containing top-secret cables, including U.S. arms reduction documents. In a remarkable twist to what initially appeared to be a stranger-than-fiction saga, on December 14, 1986, Lonetree presented himself to the CIA station chief in Vienna and confessed to his acts of diplomatic duplicity. He was immediately turned over to the Navy Intelligence Service (NIS), placed under arrest, and charged with espionage. Lonetree was convicted on multiple counts of turning over classified information, court-martialed in 1987, and sentenced to 30 years in prison.

He would infamously be the first U.S. Marine Corps member ever convicted of espionage. Because he cooperated with authorities, his sentence was reduced to 25 years, of which he served nine before being released in February 1996. Naturally, chaos ensued as the scandal began to unfold. Secretary of State George Shultz phoned the U.S. Ambassador to Thailand, Bill Brown, in the middle of the night. Brown was a former Marine who had served in Moscow and was considered a valuable resource. As Brown notes in his oral history, when the Secretary of State *"asks you to do something, you do it."* Brown went straight to Washington to help remedy the situation and smooth things over, as it were.

However, what was initially supposed to be a quick fix turned into an ordeal that would take a decade for everyone involved to overcome. Embassy Moscow was assumed to have been so infiltrated that staff took to using children's "magic slate" writing pads to pass messages back and forth. At the same time, the new embassy building was discovered to be *"one huge KGB radio station."* Sadly, it was a very low point in American security at any foreign diplomatic mission.

In the aftermath of the 1987 Lonetree scandal, Embassy Moscow expelled all Soviet domestic workers from the compound due to heightened concern about espionage activities. The consensus among embassy officials was that these workers could potentially be used by Soviet intelligence services to gather sensitive information or engage in espionage activities. Unfortunately, the expulsion of all Soviet domestic workers from the U.S. Embassy in Moscow had a significant impact on the families of diplomats stationed at the embassy. Before the expulsion, many families had relied on Soviet domestic workers for various household tasks, such as cleaning, cooking, and childcare. Following the expulsion, however, embassy staff and their families were suddenly left without the essential support and services previously provided by the Soviet domestic workers. This abrupt change resulted in families having to manage household responsibilities independently or find alternative arrangements for domestic help.

That was the rather untenable situation we encountered when we arrived in Moscow in 1988. My family had to adjust to the challenges of not having domestic help. However, we were fortunate that our children were both teenagers and quickly learned how to assist us in managing basic household tasks. Additionally, as African Americans, Annie and I had grown up in some of the most impoverished neighborhoods in America, and we never even considered hiring domestic staff during our time in Moscow. Despite the challenges of working all day and attending diplomatic receptions and formal dinners practically every day of the working week, through efficiently coordinated teamwork, we survived the entire two-year tour in Moscow without hiring domestic help.

CHAPTER TWENTY-SIX

THE DUTIES AND RESPONSIBILITIES OF AN ASSISTANT AIR ATTACHÉ

"Diplomacy is listening to what the other guy needs. Preserving your own position but listening to the other guy. You have to develop relationships with other people so when the tough times come, you can work together."
- **Colin Powell,** 65th Secretary of State of the United States

The position of Assistant Air Attaché at the Embassy in Moscow had been my lifelong ambition. It was more than a dream; it had long transformed into a passionate career aspiration by the time I stepped onto the lower rungs of the officer cadre career ladder of my beloved service branch, the United States Air Force. To finally fulfill that ambition, I had both wittingly and unwittingly engaged in higher studies and mastery of the Russian language. I had also gone to great lengths to apply for the position at strategic points in my remarkably brilliant career, and perhaps more importantly, I had finally resorted to deploying the formidable arsenal of connections and contacts I had garnered throughout my career to support my recommendation for the position. All my efforts proved eminently worthwhile, especially since I would be joined on that diplomatic tour by my beloved Annie, who seemed to appear in my life at just the right moment.

When my family and I arrived in Moscow in the summer of 1988, I was more than ready for the job. As an Assistant Air Attaché, I was tasked with various duties and responsibilities related to representing the national interest of the United States in the Soviet Union. Essentially, my job was to help build and sustain a mutually beneficial relationship between the United States and the Soviet Union. I was also required to advise the United States ambassador to the Soviet Union and the Embassy country team on the latest developments in Soviet aerospace weaponry. In other words, my primary duty as an Assistant Air Attaché was to be an "overt" observer and reporter of life and society in the Soviet Union and to write reports on the Soviet Armed Forces, especially the Soviet Air Force.

Another of my responsibilities was to serve as a military escort on special airlift missions and United States diplomatic flights, conveying the ambassador and senior US officials to, from, and within the Soviet Union. If military aircraft were used, the attaché was responsible for seeking and processing aircraft clearances and arranging aircraft servicing and aircrew accommodations in the Soviet Union. As the convention dictated, Annie and I were an attaché-wife team, which involved performing numerous representational duties while assigned to the US embassy in Moscow. For example, on average, from Monday through Friday, we attended at least one diplomatic reception and one formal dinner nearly every evening. Additionally, when we were not attending these events, we were responsible for hosting guests from the foreign diplomatic community for "sit down" dinners in our residence.

Diplomatic receptions at US embassies abroad are integral to American diplomacy, fostering relationships, promoting cultural exchange, and advancing US interests. These events provide a platform for American ambassadors, diplomats, and military attachés to engage with foreign leaders, dignitaries, and citizens, showcasing American culture, values, and hospitality. The primary purpose of diplomatic receptions at US embassies abroad is to promote bilateral relations by strengthening ties between the US and the host country, fostering cooperation and understanding. Naturally, American interests are advanced through support for US foreign policy objectives, such as trade promotion, security cooperation, and human rights advocacy. These receptions also serve as a powerful platform to showcase American culture and share American values, customs, and traditions with foreign audiences, promoting cross-cultural understanding and exchange. All of this is aimed at building relationships and networks, as diplomats and attachés maintain connections with key stakeholders, including government officials, business leaders, and civil society representatives.

US embassies abroad host various types of diplomatic receptions, including Independence Day celebrations that commemorate July 4th with events often featuring live music, food, and cultural performances. The National Day receptions celebrate the host country's national day or independence anniversary, promoting bilateral relations and cultural exchange. Other events include cultural exhibitions that showcase American art, music, literature, and film, as well as business and trade receptions that promote

American business interests and investment opportunities in the host country. There were times when Annie and I received official invitations to attend receptions or dinners hosted by the US ambassador. However, at these social events, we were not guests but working members of the ambassador's staff. Our instructions were to socialize and mingle with guests, engaging them in conversation and generally attending to their needs during the function.

Diplomatic receptions at US embassies abroad promote American diplomacy, culture, and interests. By fostering relationships, advancing American interests, and showcasing American culture, these events contribute to a more nuanced understanding of the United States and its values. As the world becomes increasingly interconnected, diplomatic receptions will continue to serve as essential tools of American diplomacy, promoting peace, prosperity, and mutual understanding. After attending various events throughout the week, Annie and I reserved Saturday and Sunday to spend quality time with Karen and Torrey. Whether we were staying home to watch a movie, playing board games, going sightseeing and shopping, or attending church services on Sunday, the weekend was always reserved for family.

One of the major responsibilities of being an attaché in Moscow was the requirement to travel throughout the Soviet Union. During my two-year tour as an attaché, with the permission of the Soviet Military Foreign Liaison Office, I traveled to ten of the sixteen Soviet Republics. This was no easy task because, in the 1988-1989 session, roughly two-thirds of the Union of Soviet Socialist Republics (USSR) was closed to foreigners. This restricted area was known as the "closed city" system, which included regions with sensitive military installations, scientific facilities, and other strategically important sites. To whatever extent possible, the Attaché Office recommended that my travel plans include my wife, Annie. However, since she had a full-time teaching position at the Anglo-American School and served as the bus monitor for elementary-aged embassy students attending classes at the Anglo-American School's campus in downtown Moscow, she was able to join me on travel missions on only two occasions during our two-year tour.

CHAPTER TWENTY-SEVEN

FAMILY ALWAYS COMES FIRST

"The Cold War isn't thawing; it is burning with a deadly heat. Communism isn't sleeping; it is, as always, plotting, scheming, working, fighting."
- **Richard M.** Nixon, 37th President of the United States

Although I had fulfilled my dream of becoming the first African American Assistant Air Attaché and diplomat at the American Embassy in Moscow, I remained conscious that I owed a significant part of my success to my wonderful family. Annie and the children had given me their unwavering support in every way possible to achieve that goal. They had further stretched themselves to accommodate the sacrifices my new office and status demanded. For all that and more, I owe them an enormous debt of gratitude that is, and will always be, difficult to express.

When we arrived in Moscow, I made it my mission to ensure that my family maintained as close a semblance as possible to their lifestyle in the United States. However, this was not an easy task.

A significant obstacle that immediately appeared was the sense of isolation that accompanied the Cold War of that era. It is pertinent to provide a brief background on the Cold War. It was essentially a political and economic rivalry between the United States and the Soviet Union that lasted from World War II until 1991. It began after World War II when the alliance between the United States, Great Britain, and the Soviet Union started to fall apart. The Soviet Union established communist governments in Eastern Europe to protect against a possible German threat.

The Cold War was fought mainly on political, economic, and propaganda fronts. Some key events during the Cold War included the *Space Race,* a competition between the United States and the Soviet Union to be the first to put a human in space, and the *Cuban Missile Crisis,* a dangerous confrontation between the two superpowers in 1962 when the Soviet Union secretly installed missiles in Cuba. The Cold War ended in 1991 when the Soviet Union dissolved into fifteen independent nations. The fall of the Berlin Wall in 1989, free elections in Eastern Europe, and economic reforms by Mikhail Gorbachev, the last leader of the Soviet Union, all contributed to its conclusion.

Not surprisingly, during the Cold War, the political climate between the United States and the Soviet Union was characterized by extreme tension, deep suspicion, and constant fear of potential conflict, fueled by ideological differences between *capitalism* and *communism*, a massive arms race, and proxy wars fought across the globe, all while the two superpowers avoided direct military confrontation with each other. This resulted in a climate of constant political maneuvering, propaganda, and espionage, where every action by one side was viewed with extreme caution and often interpreted as a threat by the other. Due to this tense political climate, security measures were strict. Karen and Torrey constantly remarked that they felt as if they were *"living in a bubble"* within the confines of the embassy compound. There was little doubt about the feeling of isolation from the local Russian community, with severe restrictions on their movements outside the embassy grounds.

Since we had arrived in Moscow during the summer, I made it my mission to accompany the family on visits to some of the city's numerous prominent sightseeing venues every weekend. We visited Red Square, the world-famous Bolshoi Ballet, Moscow Circus, Gorky Park, the Moscow Metro, the Arbat Shopping District, and the many museums and theaters that dotted the Moscow landscape. This allowed Annie, Karen, and Torrey to experience firsthand what life was like for the average Russian citizen in Moscow and compare it to their previous experiences as citizens of the United States. For our two teenagers, this was an incredibly eye-opening experience. Unfortunately, during some of our travels around Moscow, whether on the subway escalators or waiting for the metro, we often received catcalls and racial slurs from some Russian citizens. Perhaps one could not blame them; rarely had they seen a well-dressed, English-speaking African American family riding on public transportation within the city. Even so, it remains reasonable to expect that they could have met their surprise with more decorum. Yet, living in the embassy compound had its benefits. Some amenities for the use of diplomats and their families included playgrounds, a small commissary, a beauty and barber shop, a gym, a swimming pool, tennis, basketball, racquetball, squash courts, saunas, lockers, a weight room, a cafeteria, a bowling center, and a video rental center.

My next priority was to enroll Karen and Torrey at the Anglo-American School (AAS) in Moscow. The Anglo-American School of Moscow was an independent, non-profit, co-educational, international day school catering to students between the ages of 4 and 18. The New England Association of Schools and Colleges, the Council of International Schools, and the International Baccalaureate Organization fully accredited the school. It was also a member of the National Association of Independent Schools, the European Council of International Schools, and the Central and Eastern European Schools Association.

Founded in 1949, the Anglo-American School (AAS) of Moscow served the educational needs of diplomatic children who could not attend Soviet schools. The school was a private academy administered by the British, Canadian, and American governments. Students of foreign diplomatic staff represented more than 40 nationalities. While children in grades 7 through 10 attended classes on the embassy compound, elementary students were bused approximately 30 minutes to the school facility located downtown. Fortunately, since Karen and Torrey were scheduled to enter the ninth and eighth grades, they would attend all their classes on the embassy compound.

As soon as we were officially notified of our selection to go to Moscow, Annie contacted the Director of AAS to inquire about employment opportunities as an elementary school teacher. After forwarding her credentials to the school, including her current teaching license, college transcripts, and letters of recommendation, Annie received confirmation from the director that she would be hired as a kindergarten teacher. Annie was ecstatic because she would be the first African American to be employed as a teacher in the school's history. However, when we arrived in Moscow and Annie reported to AAS to begin her employment, she was confronted with a serious roadblock. The wife of an embassy employee claimed that she had been promised the vacant teaching position at AAS and that Annie did not have the required qualifications. This individual, who managed the embassy commissary, threatened to file a formal complaint against the school if she was not given the job.

Due to this unexpected roadblock, the director advised that Annie resubmit her credentials before being considered for the position. Although she was appalled and distraught by this unfortunate turn of events, Annie quickly obtained copies of her excellent credentials from officials in the United States and forwarded the requested information to AAS. To our utter surprise, we subsequently learned that the individual questioning Annie's credentials had not even received a bachelor's degree in education and had no previous experience as a teacher. She relied on patronage and her supposed contacts within the embassy to secure a position for which she had no qualifications. Although the proverbial devil is always busy, through determination and prayer, we were able to overcome this obstacle. Finally, Annie was hired as the first African American to teach at the Anglo-American School of Moscow. In her classes, she taught students from over twenty different countries.

CHAPTER TWENTY-EIGHT

TEENAGE BLUES IN A FOREIGN LAND

"We have to acknowledge that adolescence is that time of transition where we introduce children to the idea that life isn't pretty, that there are difficult things, that there are hard situations, and that it's not fair. Bad things happen to good people."
- Laurie Halse Anderson

My two teenagers, Torrey and Karen, were initially reluctant to leave their friends and family behind in Washington, DC, and travel halfway around the world to Moscow to accompany me on my attaché assignment. However, after serious conversations about the positive aspects of experiencing life in Russia, they both relented and finally told me they would support me in any way possible in fulfilling my lifelong dream of becoming an Assistant Air Attaché. But it wasn't until we arrived in Moscow that I began to understand the unique pressures my children would face, primarily due to their identity as Black American teenagers and residents of the United States Embassy compound in Moscow.

Although the family had just concluded a year in Germany before coming to the Soviet Union, adjusting to a different culture and way of life, both in Moscow and the US embassy community, posed enormous challenges for our children in terms of language, customs, and social norms. It didn't take long for me to conclude that living in the US embassy compound could, at times, be excruciatingly isolating, especially for Black teenagers who were already feeling a sense of separation from their peers and friends in the society in which they were born and raised. To compound matters, their sense of isolation was intensified by belonging to a racial minority in a predominantly White diplomatic community. In essence, theirs was a despairing case of double jeopardy. On one hand, they felt isolated even from their fellow Americans simply because they were unwitting members of a racial minority. On the other hand, they were exposed to the culture shock of being both foreigners and a racial rarity in a relatively closed society.

Soviet society was considered extremely closed in that era, with tight control over information, limited freedom of movement, severe restrictions on political dissent, and a one-party system, effectively isolating citizens from the outside world and heavily regulating their lives. In particular, the Soviet government heavily censored media, restricting access to information about political events, foreign news, and even artistic expression that could be seen as critical of the authorities. Citizens needed special permission to travel abroad, making international travel nearly impossible for most people and, by extension, heavily monitoring the movement and activities of foreigners within the country. The atmosphere of apprehension was heightened by the fact that the KGB, the Soviet intelligence agency, closely monitored the activities of both citizens and foreigners, including their political views and social interactions, creating an environment of fear. Naturally, the government controlled nearly all aspects of the economy, often leading to shortages of consumer goods and long lines in stores. It wouldn't be until the very late 1980s that President Mikhail Gorbachev introduced the reforms of *Glasnost,* or openness, and *Perestroika,* or restructuring, aimed at gradually loosening some of these restrictions to open up Soviet society.

In addition to grappling with questions about their identity and belonging that inevitably impacted their self-image, my two teenagers also faced pressures related to racial identity, diplomatic status, and cultural concerns. Although approximately ten male students from African diplomatic families were enrolled at the Anglo-American School, Torrey was the only African American on campus. Torrey was an excellent student. However, in the traditional, well-structured teaching environment of the Anglo-American School, he was considered overly loud and impulsive by some of his teachers and White classmates. Yet, my teenager was simply being a "normal African American teenager." For Torrey, this misperception meant peer pressure to conform to specific codes of behavior, attitudes, and norms, exacerbated by expectations and stereotypes about his race.

Torrey was a Black teenager coming of age during the 1980s. He would often walk around the embassy compound with his large boom box blasting music while singing the song by American rapper LL Cool J, *"I Can't Live Without My Radio."* Although Annie and I received complaints from embassy residents about our son's boisterous behavior, we ignored those complaints because Torrey was merely exercising his own cultural identity as a typical Black adolescent. Although the Anglo-American School had a robust sports program for students, including basketball, volleyball, soccer, and swimming, it did not have a flag football program. When Torrey arrived in Moscow, he organized the first-ever flag football team at the US embassy. He taught his fellow students from Nigeria, Canada, Yugoslavia, Great Britain, Spain, and Germany the fundamentals of American football. With Torrey doing double duty as both a coach and player, the AAS flag football team played matches against the Marine Security Guards, the State Department staff, and members of the French Embassy.

Building on his experience as an ice skater in Germany, Torrey learned how to play ice hockey in Moscow and even scored goals against some of his opponents. Living in a diplomatic compound in the Soviet Union during the Cold War came with security concerns and movement restrictions for the children of diplomats. In one instance, Torrey and six classmates left the embassy compound alone and took a local bus to Red Square to sightsee. Unfortunately, the group lost its way in Moscow and did not know how to return to the US embassy. In less than ten minutes, however, a private bus arrived to transport the group back to the embassy grounds. Yet, it was clear that the Soviet security agents responsible for protecting diplomatic staff and their children did not like the idea of a group of foreign teenagers, with little or no fluency in Russian, roaming the streets of Moscow on their own.

Because Torrey was your classic "free spirit" and a typical African American teenager, he became a constant target of bullying and name-calling by some of his White classmates. Whether on the playground or in the boys' restroom, these students called Torrey names such as *"Bama," "Burnt Toast,"* and "Monkey." Torrey never told us about the bullying he experienced in school. He merely suffered in silence. However, one day, at the end of classes, the group's ringleader, the son of a high-ranking embassy official, began to harass Torrey on the playground again. That was the breaking point for my son, and he and his harasser almost came to blows. Fortunately, the other students intervened and quickly stopped the altercation. Even though this incident occurred on the embassy grounds after school hours, the AAS principal suspended only Torrey for creating a disturbance.

In contrast, none of the White students involved in the incident received any punishment. When Annie learned about Torrey's suspension, she was livid. First, she contacted the school principal to express her displeasure about how Torrey was treated. She then demanded that the principal convene a hearing with all the students involved in the unfortunate situation. During the hearing, Annie got all of the perpetrators to acknowledge that they had been bullying and harassing Torrey since the first day of school. After the hearing, Torrey was fully exonerated and released from suspension. This entire event made me reminisce about the times in my childhood when I was harassed and discriminated against simply because of the color of my skin.

Fortunately, our daughter Karen did not experience the same level of racial discrimination or prejudice as her brother Torrey while attending the Anglo-American School of Moscow. Because of her bright and outgoing personality, she was warmly welcomed at the school by students and faculty members alike. Also, Karen quickly made friends with her female classmates from various African countries. Karen's friendships and interactions with these students gave her a sense of identity, belonging, and sisterhood within the school community. Additionally, because many of her African classmates were not proficient in English, Karen spent much time and effort learning French and Russian in Moscow to communicate effectively with her fellow students and friends from the African continent.

My two teenagers' years living in Moscow greatly impacted their future lives. Torrey graduated from Virginia State University with a degree in Sports Management and is now a secondary school mathematics teacher and football coach. After returning home from Moscow, in her senior year of high school, Karen attended the prestigious summer program in Russian Studies at the Virginia Governor's School hosted by James Madison University. She then received her bachelor's and master's degrees from James Madison University and a PhD in Public Policy from Walden University. Karen and Torrey often reminisce about their experiences while living at the US embassy in Moscow. Over the years, our children have regularly attended class reunions sponsored by their former classmates at the Anglo-American School of Moscow.

CHAPTER TWENTY-NINE

NEVER SAY "NO" TO THE FIRST LADY

"Giving frees us from the familiar territory of our own needs by opening our mind to the unexplained worlds occupied by the needs of others."
- **Barbara Bush,** First Lady of the United States (1989 to 1993)

On December 7, 1988, a massive earthquake measuring 6.9 on the Richter scale struck the Soviet Republic of Armenia, one of the constituent republics of the Soviet Union and an area known to be vulnerable to large and destructive earthquakes, part of a larger active seismic belt that stretches from the Alps to the Himalayas. The epicenter of the tremor was near the town of Spitak, located approximately 43 miles northwest of Yerevan, the capital city of Armenia. The earthquake resulted in widespread destruction and loss of life. An estimated 25,000 to 50,000 people were killed, while roughly 130,000 others were injured and displaced. Additionally, the quake caused significant damage to infrastructure, including buildings, roads, and utilities. The immediate response to the earthquake faced challenges due to the enormous scale of destruction and harsh winter conditions.

The damage was significant, primarily because some of the most intense shaking occurred in industrial areas, such as chemical and food processing plants, electrical substations, and power plants. The Metsamor Nuclear Power Plant, about 47 miles from the epicenter, experienced only minor shaking, and no damage occurred there, but it was closed for six years due to vulnerability concerns. It reopened in 1995 to relieve Armenia of a significant energy blockade imposed by Azerbaijan. The situation was so critical that, despite the tensions of the Cold War, Soviet leader Mikhail Gorbachev felt compelled to formally ask the United States for humanitarian help within a few days of the earthquake, the first such request since the late 1940s. The United States government responded by providing substantial aid and assistance to the disaster victims. In total, one hundred countries and numerous civilian relief organizations sent significant humanitarian assistance to Armenia in the form of rescue equipment, search teams, and medical supplies. Private donations and help from non-governmental organizations also made up much of the international aid effort. While transporting some of these supplies to the region, a Soviet aircraft carrying nine crew members and 69 military personnel, and a transport plane from Yugoslavia were both destroyed in separate incidents. As an Assistant Air Attaché, the US Embassy in Moscow assigned me to Armenia as part of the American humanitarian relief effort. I served as the Air Force on-site Mission Commander and coordinated a total of 54 sorties by US Air Force transport aircraft, bringing in humanitarian supplies to the citizens of Armenia.

Significantly, in response to the earthquake, the First Lady of the United States, Mrs. Barbara Bush, wife of the 41st President of the United States, George H.W. Bush, and the nonprofit organization *Project Hope* collaborated to provide medical supplies, equipment, and personnel to assist with the relief efforts in Armenia. As a follow-up to these humanitarian efforts, in February of 1989, First Lady Barbara Bush visited Armenia to assess the region's ongoing relief and recovery efforts and to show support for the disaster victims. As the wife of U.S. President George H. W. Bush, her visit helped to highlight the importance of international aid and solidarity in times of crisis. While in Armenia, the First Lady met with survivors, relief workers, and local officials to understand the impact of the earthquake and offer compassion and support.

Before returning to Washington, Mrs. Bush organized the airlift of 35 severely injured Armenian children to the United States for medical treatment. I had the distinct, if not unique, privilege of coordinating the airlift for Mrs. Bush. Before she departed from Yerevan airport, I wished her well on the long flight back to Andrews AFB near Washington, DC. However, before boarding her plane, the First Lady turned to me and said, *"Colonel, I want you to join us on the flight back to Andrews Air Force Base. Because you speak fluent Russian, you can communicate with and comfort the injured children on the long flight back to the United States."* Naturally, I was taken by surprise at Mrs. Bush's request. My immediate concern was that I had no authorization from the Defense Attaché Office or the Embassy to make an unauthorized trip back to Washington, DC. My travel orders specifically stated that I must return to Moscow after completing my work in Armenia. Nevertheless, I instantly resolved that I would be the last person to refuse a direct order from the wife of the President of the United States. My response to America's First Lady was, *"Yes, Mrs. Bush, it will be my pleasure to accompany you and the children back to Washington."*

When the aircraft landed at Andrews AFB, representatives from hospitals throughout the United States were waiting to transport the injured Armenian children for medical treatment. Immediately, I exited the aircraft and headed straight for the passenger terminal to call the Defense Attaché in Moscow to inform him of my unauthorized trip to Washington on board the First Lady's aircraft, and that I needed travel orders to purchase an airline ticket to return to Moscow. After a few moments of silence, the general replied that I should book my return trip to Moscow as soon as possible. As luck would have it, I had the opportunity to visit my mother in Washington, DC, for several days before returning to Moscow.

My encounter with First Lady Barbara Bush will forever remain a cherished moment in my career as an officer of the United States Air Force and a diplomatic representative of our great nation abroad. In a life deeply embedded in America's politics, Mrs. Bush remained a largely apolitical figure. She exuded cultural and emotional appeal across partisan divides. Her wry, down-to-earth view of herself and the world emphasized her common humanity and respect for others. Although Mrs. Bush became well known for her advocacy of domestic causes, including literacy and civil rights, what is less remembered by many Americans is that Barbara Bush played significant roles in some of the most critical American diplomatic and peacemaking challenges of her lifetime.

As first lady, she played a significant role in defining US-Soviet relations. Apart from her heroic efforts in the wake of the Armenian earthquake of 1988, another iconic moment was her

1990 joint appearance with Soviet First Lady Raisa Gorbachev to deliver the commencement address at Wellesley College, a private women's liberal arts college in Wellesley, Massachusetts, and one of the nation's premier universities. In that shared act, the two women broke traditional molds for female leaders in the world's superpower nations. They jointly clarified that our only future is dialogue and building relationships. The event was powerful and emotional as Mrs. Bush and Mrs. Gorbachev evoked standing ovations and garnered global news coverage. Barbara Bush's 1990 partnership with Raisa Gorbachev was unequivocally an illustration of her role as one of the most widely respected women of her time and her willingness to take a stand to make a difference.

CHAPTER THIRTY

MILITARY TO MILITARY EXCHANGE

"During my tour of duty in Russia from early 1942 to late 1943, three kinds of aid were coming to Soviet Russia from our country. First was Lend-Lease, in which I was especially interested officially. Second, shiploads of supplies were being delivered to Russia by the American Red Cross, supposedly for the relief of exiled Poles and needy Russians displaced by the war. Third, Russian Relief Supplies were being gathered and sent to Russia by Russian Relief Societies all over the United States."
- **Admiral W. H. Standley,** USN (Ret.), US World War II Ambassador to the U.S.S.R.

The year was 1988. I was assigned by the US Defense Attaché in Moscow to escort the first group of US military physicians visiting their Soviet counterparts as part of a military exchange agreement between the United States and the Soviet Union. The exchange aimed to foster collaboration and mutual understanding between the military and medical professionals of the two countries. Being selected by the Defense Attaché Office as escort and interpreter for this group of senior military doctors on their inaugural visit to the Soviet Union was quite an honor. However, in preparation for this visit, I had to spend many hours studying a Russian-to-English dictionary to familiarize myself with medical terminology.

US-Soviet military-to-military exchanges in the 1990s significantly developed the relationship between the two superpowers. After the end of the Cold War, the United States and the Soviet Union initiated a series of military exchanges to build trust, promote transparency, and reduce tensions. Notable examples of such exchanges included the US-Soviet Military Maritime Consultative Group (MMC), established in 1990, which focused on preventing incidents at sea and promoting safe navigation practices. The US-Soviet/Russian Defense Conversion Committee was established later, in 1992, to explore ways to convert military industries to civilian use. In the same year, the US-Russia Joint Commission on POW/MIAs was established to account for missing American service members from World War II, the Korean War, and the Vietnam War. Again, in that same year, the Nunn-Lugar Cooperative Threat Reduction (CTR) program was launched to provide US assistance to Russia and other former Soviet states in dismantling and securing their nuclear weapons and materials. Suffice it to say that these exchanges played a crucial role in fostering cooperation and understanding between the US and Soviet/Russian militaries. They helped lay the groundwork for future collaborations on counterterrorism, non-proliferation, and regional security.

More specifically, US-Soviet military medical exchanges and collaboration refer to the historical interactions and cooperation between the medical communities of the United States and the Soviet Union, particularly during the Cold War and post-

Cold War eras. Although the two superpowers were ideological adversaries, they engaged in various medical exchange programs and collaborative efforts, driven by mutual interests in advancing medical science, improving healthcare, and reducing the risk of nuclear war. Notable examples of US-Soviet military medical exchanges and collaboration included the *US-USSR Health The Agreement of* 1972 facilitated cooperation in various medical fields, including cardiology, oncology, and infectious diseases. Medical exchange programs were also organized in which American and Soviet medical professionals participated, visiting each other's medical facilities, attending conferences, and sharing knowledge on medical research and practices. I participated, in my capacity as Assistant Air Attaché, as a diplomatic escort and interpreter during one such exchange.

The United States and the Soviet Union also collaborated on disaster response and emergency medicine, including joint exercises, training programs, and research related to nuclear medicine, radiation protection, and the medical consequences of nuclear war. Both countries actively participate in international health organizations, such as the World Health Organization (WHO), to address global health issues and promote cooperation. Naturally, these exchanges and collaborations have advanced medical knowledge and helped build bridges between the two nations, fostering mutual understanding and respect.

During the seven-day visit of US military physicians, the US Deputy Surgeon General, an Air Force brigadier general and four military physicians had the opportunity to meet with senior members of the Soviet Ministry of Defense Medical Staff to observe and learn from their Soviet counterparts in various medical settings, including military hospitals and clinics. They also participated in discussions and academic exchanges to share knowledge and best practices in military medicine. The exchange provided valuable insights into the medical care and treatment approaches of both countries. It helped build bridges between the military medical communities of the United States and the Soviet Union during the Cold War era, marking a significant step toward enhancing cooperation and collaboration in military medicine.

Naturally, the visit also helped foster personal relationships and build connections between individual service members. These interactions were crucial in promoting mutual respect and understanding, paving the way for future cooperation and collaboration between the US and Soviet militaries.

CHAPTER THIRTY-ONE

THE AMERICAN CONGRESSIONAL DELEGATION TO ARMENIA

"The limits of my language are the limits of my world,"
- A famous quote attributed to Austrian philosopher **Ludwig Wittgenstein**

In 1989, US Representative Robert Torricelli led a congressional delegation (CODEL) to The Soviet Union. The delegation was part of US efforts to improve international relations with the Soviet regime during a tumultuous period marked by the end of the Cold War. A *CODEL,* an acronym for congressional delegation, is a group of members of the United States Congress who travel together on official business. These trips are often referred to as "codels." They can include visits to other countries, meetings with foreign officials, or inspections of military or diplomatic facilities.

Codels are standard practice in Congress and serve many purposes. They can be used for fact-finding missions and to gather information and insights to inform legislation or policy decisions. Many aspects of Washington, at least on the strictly official level, can appear mystifyingly bureaucratic to the voters who send lawmakers there, and in many respects, codels are no exception. Yet, there is a perfectly valid reason why such trips are referred to as a *"secret weapon"* in what is famously considered a gridlocked capital. For more than half a century, visits intended to reassure America's allies about the goings-on in the US have also helped members of Congress foster rare human connections that have influenced future American policy on the global stage and even on issues unrelated to foreign affairs.

On paper, codels allow lawmakers to travel abroad to meet with world leaders, diplomats, and advocates on national security topics. But in practice, lawmakers who join them spend significantly more time together than they do on the Hill each week while running between committees, staff meetings, and votes. The rigid, often scripted nature of their typical day essentially disappears, allowing codels to counterbalance domestic polarization. Codels are also frequently used for diplomacy, building relationships with foreign governments, and advancing the United States' interests. In addition to the members of Congress, codels often include staff members and other officials who can provide expertise or support on the issues being addressed. For example, a codel might consist of staff members with expertise in foreign policy or defense or officials from relevant agencies such as the State Department or the Department of Defense.

Codels are typically organized and funded by the relevant committees in Congress, such as the Foreign Affairs Committee or the Armed Services Committee. These committees often set the trip's agenda, determine who will be included in the delegation, and provide logistical support. Codels can be controversial, as they may be seen as a way for members of Congress to take taxpayer-funded trips to exotic locations. Critics have also raised concerns about potential conflicts of interest, such as when members of Congress are invited on codels by foreign governments or other interested parties.

As Assistant Air Attaché, I was actively involved in the 1989 congressional delegation (CODEL) led by US Representative Robert Torricelli to the Soviet Union. During that visit, Representative Torricelli, a Democrat who served in the US House of Representatives from New Jersey's 9th district from 1983 to 1997 and as a United States Senator from New Jersey from 1997 to 2003, and his delegation met with Soviet officials to discuss a wide range of issues, including arms control, human rights, and bilateral trade. After concluding their business in Moscow, the delegation wanted to visit several Soviet cities to meet with local leaders and gain firsthand insights into daily life in the country.

As an Assistant Air Attaché, I was assigned to serve as the US military air escort for the delegation's trip from Moscow to Yerevan, Armenia. Yerevan is the capital and largest city of Armenia and one of the world's oldest inhabited cities. Situated along the Hrazdan River, Yerevan is the administrative, cultural, and industrial center of Armenia. As noted previously, per US and Soviet protocol, US military aircraft traveling in and out of Soviet airspace must have on board a Soviet navigator and a Russian-speaking Air Force Attaché to ensure flight safety. Unfortunately, the Soviet navigator missed or was omitted from this particular flight, and I was the only Russian-speaking individual on board the aircraft for the flight to Yerevan.

When we approached the airfield in Yerevan, the pilot asked me to join him in the cockpit. He was having great difficulty understanding the approach and landing instructions from the air traffic control tower. Under regulations, all Soviet air traffic controllers must be proficient in English. However, in this case, the Armenian air controllers seemed to have very limited fluency in the English language. I immediately contacted the tower in Russian and was informed that the airport was closed due to blowing snow and strong crosswinds. The tower instructed me to tell the Air Force pilot to abort the landing and head for an alternative airfield in Russia.

Contrary to that instruction, however, Representative Torricelli told me emphatically that he had no desire to fly to an alternative airfield in Russia. He asked me to inform the tower that we requested a heading to fly to Turkey. A few hours later, despite having extremely low fuel, we safely landed in Ankara, Turkey. This is one example of how my Russian language training paid off in a crisis. If the pilot had attempted to land on a closed runway at Yerevan Airport during a snowstorm, everyone on board might have perished.

CHAPTER THIRTY-TWO

A MEETING WITH CONDOLEEZZA RICE

"The essence of America; that which really unites us, is not ethnicity, or nationality or religion; it is an idea, and what an idea it is. That you can come from humble circumstances and do great things."
- Condoleezza Rice

In late July and early August of 1989, in the twilight months of the Soviet Union, United States Secretary of State James Baker visited Moscow. Secretary Baker's visit was part of diplomatic efforts to improve relations between the United States and the Soviet Union during the waning years of the Cold War. Another purpose of the visit was to lay the groundwork for a proposed summit between President George H.W. Bush and Soviet President Mikhail Gorbachev, scheduled in Washington, DC, in May 1990. A key member of the secretary's delegation was Dr. Condoleezza Rice, who was serving on President Bush's National Security Council staff as Director of Soviet and East European Affairs.

Dr. Rice was the quintessential diplomat. Born on November 14, 1954, in Birmingham, Alabama, Condi, as she was affectionately known in diplomatic circles, earned a Bachelor of Arts in political science in 1974 from the University of Denver, a master's degree in political science from the University of Notre Dame in 1975, and a doctorate in political science from the Graduate School of International Studies at the University of Denver in 1981. This marked the beginning of a spectacular career in academia, public service, and diplomacy. After a stint at Stanford University as a political science professor, in 1987, she served as an advisor to the Joint Chiefs of Staff and, in 1989, was appointed director of Soviet and East European Affairs on the National Security Council. Following another tenure at Stanford University as provost, in 2001, she was appointed National Security Advisor by President George W. Bush. She succeeded Colin Powell as Secretary of State in 2005, becoming the first African American woman to hold the position.

As the 66th United States Secretary of State, Dr. Rice supported the expansion of democratic governments and championed the concept of *"Transformational Diplomacy,"* which aimed to redistribute US diplomats to areas of severe social and political turmoil, address issues such as disease, drug smuggling, and human trafficking, and emphasize aid through the creation of the position of Director of Foreign Assistance. She played a key role in negotiating several agreements in the Middle East, including the Israeli withdrawal from and the opening of the Gaza border crossings in 2005 and the August 14, 2006, ceasefire between Israel and Hezbollah forces in Lebanon. Dr. Rice organized the Annapolis Conference on November 27, 2007, which focused on finding a two-state solution to the Israeli-Palestinian conflict. She also worked actively to improve human rights issues in Iran and supported the passage of a United Nations Security Council Resolution imposing sanctions against the country unless its uranium enrichment program was curtailed.

Another primary concern for Dr. Rice was North Korea's nuclear program and its subsequent testing of a nuclear weapon. She stood firmly against holding bilateral talks with North Korea but welcomed their participation in the Six-Party Talks among China, Japan, Russia, North Korea, South Korea, and the United States. In October 2008, one of her most successful negotiations came to fruition with the signing of the *US-India Agreement for Cooperation Concerning Peaceful Uses of Nuclear Energy (123 Agreement),* which allowed civil nuclear trade between the two countries.

I met Dr. Rice under rather fortuitous circumstances. One day, while I was conducting some business at *Spaso House,* the residence of the US Ambassador to the Soviet Union, I caught sight of a tall, well-dressed African American lady walking down the hall.

Noticing that I was also African American, she approached me, and we started a conversation. I introduced myself as Lieutenant Colonel Lewis Wallace, an Assistant Air Attaché at the embassy. She replied that her name was Dr. Condoleezza Rice, or "Condi," and that she was residing at the ambassador's residence while attending a series of high-level meetings at the Kremlin. What she said next truly surprised me: *"Colonel Wallace, my trip to Moscow has been very successful. However, now I would like to go somewhere to relax."*

Immediately, I invited Dr. Rice to be Annie's and my house guest at our quarters on the embassy compound. She accepted my invitation and stayed with us for five days before departing for Washington, DC. During discussions after enjoying one of Annie's delicious meals, Condi shared some fascinating details about her background. That was when I learned that she had been born in Birmingham, Alabama, and had grown up during the era of racial segregation in the American South. There were some interesting similarities in our backgrounds. For instance, we shared the same birthday on November 14. After completing her bachelor's degree in political science from the University of Denver, she was inducted into the prestigious *Phi Beta Kappa Academic* Honor Society. She was surprised to learn that I, too, was inducted into Phi Beta Kappa after completing my bachelor's degree in Russian Studies at Syracuse University. Finally, both Condi and I were fluent in Russian.

While she stayed with us in Moscow, we would go to the Arbat shopping district near the US Embassy daily. The Arbat was lined with various shops, cafes, and street vendors selling traditional Russian souvenirs, artwork, and handmade crafts. While shopping, we received numerous stares and looks of astonishment from Russian locals, who seemed amazed that two African Americans who spoke fluent Russian were shopping at a Russian mall. Condi was also an accomplished pianist. During her stay at our residence, we were treated, on many occasions, to her beautiful renditions on the upright piano in our quarters.

After Dr. Rice returned to Washington, Annie and I remained in touch with this phenomenal woman over the years. However, I recall an exceedingly disturbing incident involving Dr. Rice in 1990. While waiting in a line of dignitaries to bid farewell to Soviet President Mikhail Gorbachev after his visit to the United States for a presidential summit, a U.S. Secret Service agent, mistaking Condi for a mere office secretary, reportedly shoved her aside and did not allow her entry into the official party. Other White House staff members immediately came to her aid, intervening so she could take her rightful place in the farewell line. In 2005, after Dr.

Rice became the first African American woman to serve as U.S. Secretary of State, she invited Annie and me to her first *Secretary's Reception at* the Diplomatic Rooms of the State Department. It was a poignant moment, as I recalled how, many years earlier, as a lowly State Department employee working in the mailroom, I had told my mother that someday I would be invited to a fabulous diplomatic reception at the State Department. Dreams do come true.

CHAPTER THIRTY-THREE

THE MALTA SUMMIT

"I am especially glad we had this meeting. I am convinced that a cooperative U.S.-Soviet relationship can indeed make the future safer and brighter. Now, with reform underway in the Soviet Union, we stand on the threshold of a brand new era of U.S.-Soviet relations, and it is within our grasp to contribute in our way to overcoming the division of Europe and ending the military confrontation there."
- **George H.W. Bush,** 41st President of the United States, Malta, December 3, 1989

One Saturday morning in late October of 1989, I unexpectedly found myself the highest-ranking attaché at the Defense Attaché Office (DAO) in Moscow. This rather significant honor came about because the defense attaché and the principal military attachés were on scheduled travel trips to various cities across the Soviet Union. While sitting in my office that morning, the weight of my temporary status was not lost on me. I began to daydream about the day I would be promoted to the esteemed rank of brigadier general and return to Moscow as the Defense Attaché.

Perhaps because I had finally managed to accomplish my life ambition of becoming an air attaché, I found myself believing that it is essential to keep one's dreams alive.

Keeping your dreams alive is essential for a fulfilling and purpose-driven life. In my opinion, there are seven key reasons for this assertion. The first is that dreams provide a sense of direction and motivation, inspiring you to work towards achieving your goals. The second is that pursuing your dreams helps you hone your skills, build confidence, and discover your strengths and passions. The third is that achieving your dreams brings joy, satisfaction, and contentment, leading to a happier and more balanced life. The fourth is that holding onto your dreams helps you develop resilience and perseverance, which are essential for overcoming obstacles and setbacks. The fifth is that your dreams give you a sense of purpose, helping you stay focused and driven, even in the face of challenges and uncertainties. Sixth, it is the faithful pursuit of your dreams that leads to innovative ideas and creative solutions that set you apart from the crowd, drawing the attention of your superiors. Naturally, this is because the dreamer in you is also someone who can think outside the box and explore new possibilities. Finally, when you keep your dreams alive, you become a positive role model for others, inspiring them to pursue their passions and aspirations.

I had envisioned a quiet weekend at the Defense Attaché Office. I was grossly mistaken. At approximately 5:00 p.m., the office secretary informed me there was an urgent call from the White House in Washington, DC. I immediately picked up the telephone and introduced myself as the assistant air attaché on duty. The individual on the other end of the call introduced himself as a "Special Assistant to the President of the United States." The purpose of the call was to request that an air attaché from Embassy

Moscow immediately book a flight to Rome, Italy, to rendezvous with an advance team of security and communications specialists from the Office of the President. The advance team would arrive in Rome on a United States Air Force Special Air Mission aircraft.

My mission was to join the group and serve as their Air Escort and Russian interpreter for a subsequent flight to an undisclosed destination within the Soviet Union. When the aircraft arrived at that Soviet destination, we were to rendezvous with Russian security and technical specialists who would board our plane for the follow-on flight to Valletta, Malta. Upon arrival in Valletta, both US and Soviet experts would jointly inspect the Soviet cruise ship, *"Maksim Gorky,"* which was anchored at Valletta's Grand Harbor, in preparation for the planned Malta Summit on board the vessel between US President George H.W. Bush and Soviet leader Mikhail Gorbachev, scheduled for December 1989.

The Malta Summit took place on December 2 and 3, 1989. The event marked the first face-to-face meeting between the two leaders since Gorbachev assumed power in the Soviet Union. The meeting was informal and symbolic, with both leaders discussing the end of the Cold War and the possibilities for improved relations between their countries. The summit also helped lay the groundwork for future arms control agreements and improved diplomatic relations between the United States and the Soviet Union. The choice of Malta for the historic summit was, in itself, highly symbolic. The Maltese Islands are strategically located at the geographic center of the Mediterranean Sea, where East meets West and North meets South. This was significant in the context of the political and ideological divisions between the capitalist West and the Communist East.

Furthermore, Malta's choice was ideal due to the country's stance of neutrality. Malta declared its neutrality from the two superpowers in 1980, following the departure of British forces from the islands. At the time, *Neutrality was entrenched in the Constitution of Malta, which stated, "Malta is a neutral state actively pursuing peace, security, and social progress among all nations by adhering to a policy of non-alignment and refusing to participate in any military alliance."* The endeavor was also in line with Malta's views about the world, as the islands adopted a policy openly advocating against the possession of nuclear weapons. Just a year before the Summit, Malta had refused entry to a Royal Navy ship visiting the islands, citing that it would violate Malta's neutrality clause and non-nuclear policy, as it could not be confirmed whether the vessel contained nuclear weapons.

Some historians now consider the Malta Summit to be the most important meeting between the USA and the USSR since the Yalta Conference of 1945, when Franklin D. Roosevelt, Winston Churchill, and Joseph Stalin met to divide spheres of influence after the end of World War II. The Malta Summit heralded a new era of international relations and significantly reduced the immediate nuclear threat posed by the Cold War to humanity. The US and USSR leaders declared a planned reduction in troops in Europe and stated that a decrease in weaponry would be discussed at a meeting scheduled for June 1990. The desire to sign an agreement on strategic nuclear arms reduction and to move towards a chemical weapons treaty was signaled by both parties. However, the end of the Cold War ushered in an era of globalization and renewed hope for peaceful and joint progress. Regrettably, the positive spirit that characterized the post-Cold War era did not have a lasting effect. That is why the 1989 Malta Summit served as a timely reminder of the ability of states to rise to the occasion and find the necessary willpower to strive for a more peaceful world. I was honored to have played a role in the strategic and operational presentations leading to that historic meeting between two of the world's most incredible superpowers.

CHAPTER THIRTY-FOUR

THE "BLUESUITERS" AT EMBASSY MOSCOW

"The roles of the defense attaché positions in American embassies worldwide are the foundation and frontline for the military-to-military relationships that exist between the United States and other countries. We represent the United States. We are ambassadors in blue, and the most important aspect of the assignment is building relationships."

- **Brigadier General Ralph Jodice II,** Defense Attaché, US Embassy, Beijing, China (2004–2007)

Sometime in early 1989, Brigadier General Michael McRaney, Director of Public Affairs at the Office of the Secretary of the Air Force, contacted Brigadier General Ervin Rokke, the US Defense Attaché at the Embassy in Moscow to request country clearance for a two-person team of Air Force Public Affairs Specialists to visit Moscow to obtain interviews and photographs documenting the challenges faced by United States Air Force (USAF) personnel living and working in the Soviet Union. This would be the first-ever visit to the Soviet Union by a team of

Public Affairs writers and photographers. It was my distinct honor to be assigned as the primary escort and interpreter to the team on this historic visit.

The team, consisting of an Air Force staff writer and a photojournalist, departed Frankfurt International Airport on Monday, May 1, 1989, and was scheduled to arrive at Sheremetyevo International Airport in Moscow at noon. I arrived at the airport an hour before the aircraft to facilitate the team's smooth entry through Soviet Customs and Immigration. They arrived in Moscow on May Day, also known as International Workers' Day, one of the largest and most popular celebrations in the Soviet Union. It is a public holiday traditionally marked by parades, marches, and demonstrations organized by trade unions and political groups to celebrate workers' rights and achievements. May Day in the Soviet Union was a significant holiday celebrating the working class, the Soviet military, and the start of spring. The Soviet Union used May Day for international propaganda, especially during the Cold War, and the day's parades were a way to showcase Soviet military strength and, perhaps, intimidate Western countries. The Soviet Union's May Day celebrations were often held in the capital city, with the president and party secretary general in attendance. In today's Russia, May Day is officially called *"The Day of Spring and Labor."*

When I finally met our visitors at the Immigration and Border Guard Station, to my utter surprise, I learned that these two Air Force members had been allowed to board their aircraft and depart Frankfurt, Germany, without the required visas to enter the Soviet Union. Not surprisingly, upon arrival at Sheremetyevo Airport, they were immediately detained and denied entry into the Soviet Union. As a USAF Assistant Air Attaché and representative of the American Embassy, I had to utilize all the resources at my disposal and summon both persuasive force and my fluency in the Russian language to convince the KGB.

Captain of the Border Guards is to allow our visitors temporary entry into the Soviet Union without the required visas. Finally, after about thirty minutes of intense negotiations, it was agreed that our visitors would be allowed entry into the country on my recognizance. The American Embassy would contact the Soviet Ministry of Foreign Affairs immediately after the May Day holiday to request official visas for our two military guests.

The Air Force writer-photographer team did an excellent job documenting the daily challenges faced by USAF "Bluesuiters" in the Soviet Union. United States Air Force personnel are often called "Blues" or "Bluesuiters" because of the color of their service uniform. The final story, published in "Airman Magazine" in late spring 1990, helped readers understand and appreciate the sacrifice and dedication of the United States Air Force community in Moscow. Air Force personnel and their families encountered various challenges during their assignment in Moscow, and in 1989, these challenges included:

1. *Language Barriers:* Communicating with the local population and navigating daily life in a country where English is not widely spoken.

2. *Cultural Differences:* Adapting to a different culture, customs, and way of life was challenging for Air Force members and their families.

3. *Security Concerns:* Given the tense political climate of the Cold War era, there were constant concerns about security and espionage.

4. *Isolation:* Being stationed in a hostile foreign country, far from friends and family, sometimes led to feelings of isolation and homesickness.

Yet, it must be admitted that there were some positive aspects to being stationed in Moscow in 1989. They included:

1. Living in the Soviet Union gave Air Force personnel and their families a unique opportunity to immerse themselves in a different culture and better understand the world.

2. Being stationed in Moscow allowed Air Force personnel to work on unique assignments and enhance their promotion potential.

3. Being stationed in a strategic location like Moscow allowed service members and their families to travel and explore other parts of Europe and Asia.

4. From a historical perspective, being stationed in Moscow during the late 1980s placed Air Force "Bluesuiters" at a pivotal point in history, as it was a time of significant political and social upheaval in the Soviet Union.

In truth, the efforts of USAF personnel in Moscow during that era, and indeed in all US foreign missions scattered across the globe, deserve sincere commendation. As a key military diplomatic player of that time, it would be remiss of me not to use my memoirs to express such appreciation. The geopolitical landscape was fraught with tensions stemming from divergent interests, incompatible ideologies, and a renewed focus on inter-state competition for power and influence. While economic, environmental, and technological challenges were significant, security issues dominated foreign policy and global affairs. In this strategic environment, USAF personnel in Moscow were expected to leverage Soviet relationships to reduce instability and position the United States for strategic advantage. They faced the unique challenge of combining traditional and military diplomacy through bilateral negotiations to seek equitable solutions to conflicts and conduct intelligence diplomacy to spur collective action in preparation for possible defense.

Furthermore, since many bilateral and multilateral issues dealt with defense or security concerns, Air Force personnel were inevitably involved in diplomacy well before considering military action. They had to employ various non-violent measures to influence the decisions and behavior of the Soviet government and non-state actors in pursuit of America's objectives. When directed externally, these actions constituted military diplomacy and were quite significant. Such activities included dialogue, negotiations over conflict resolution or other security issues, defense and security cooperation, mild demonstrations of force, and benevolent missions like civil affairs or humanitarian assistance and disaster relief.

On June 6, 1989, Brigadier General McRaney wrote a letter of appreciation to General Rokke for the extraordinary support and cooperation of the Defense Attaché Office in Moscow during the visit by the Air Force Public Affairs team. In his letter, General McRaney wrote, *"Lieutenant Colonel and Mrs. Lewis Wallace gave unselfishly of their time to ensure our 'Airman' team obtained the interviews and photographs needed to document our Moscow story, which is scheduled to be published this fall. I especially appreciate Colonel Wallace's efforts in gaining entry for our writer and photographer when problems arose with their visas."* The General's comments are another example of the outstanding work performed by an Assistant Air Attaché in Moscow. My incredible saga continues.

CHAPTER THIRTY-FIVE

MCDONALD'S COMES TO MOSCOW

"If you can't go to America, come to McDonald's in Moscow."

- Advertising slogan for the opening of the first McDonald's in Moscow in 1990.

The first McDonald's restaurant in Moscow opened on January 31, 1990. This event, initiated and orchestrated by George Cohon, the founder and CEO of McDonald's Canada, marked the first presence of McDonald's in the Soviet Union and a significant moment in history. The opening of McDonald's in downtown Moscow, just off Pushkin Square, was a highly anticipated occasion, with thousands of Russians queuing up in the frigid cold at four o'clock in the morning to savor, for the first time, a taste of the Big Mac, French fries, and a drink. At the prohibitive price of the equivalent of half a day's wages, the cost of a McDonald's meal was relatively high for the average Soviet citizen, forcing many families to divide their Big Mac hamburgers into quarters to share. Nonetheless, it was significant that the new McDonald's set a record for serving over 30,000 people on opening day.

McDonald's, the world-famous American fast-food chain, is the largest fast food restaurant chain by number of locations, serving over 69 million customers daily in over 100 countries across more than 40,000 outlets as of 2021. McDonald's is best known for its hamburgers, cheeseburgers, and French fries, although its menu also includes other items like chicken, fish, fruit, and salads. Their bestselling item is the French fries, followed by the famous Big Mac. The McDonald's Corporation's revenues come from rent, royalties, and fees paid by its franchisees and sales in company-operated restaurants. By 2018, McDonald's had become the world's second-largest private employer with 1.7 million employees, most of whom worked at the restaurant franchises. McDonald's, founded in 1940 as a restaurant operated by Richard and Maurice MacDonald in San Bernardino, California, was transformed into a hamburger stand and later franchised. In 1955, businessman Ray Kroc joined the company as a franchise agent and, in 1961, bought out the McDonald brothers. Previously headquartered in Oak Brook, Illinois, it moved to nearby Chicago in June 2018.

The day before the official opening of the Moscow McDonald's, the president of McDonald's Canada invited selected members of the American diplomatic community at the American Embassy to attend a pre-opening reception at the new restaurant. Annie and my two teenagers were ecstatic about attending this historic event. In the past, I'd had to travel by commercial aircraft with my family to Helsinki, Finland, just to satisfy their cravings for the Big Mac. At the pre-opening reception, guests received a McDonald's Moscow T-shirt, a souvenir wristwatch, and a commemorative pin from the president of McDonald's Canada. To this day, Annie, Karen, and Torrey treasure the unique gifts they received at this special event.

On opening day, January 31, 1990, a group of teenagers from the American Embassy, brandishing their diplomatic credentials, attempted to jump the queue and enter the new restaurant ahead of the thousands of Russians waiting in line. However, as soon as they received catcalls and jeers from those in line, they quickly retreated back to the embassy. The following day, another group of American teenagers from the embassy, including Karen and Torrey, successfully entered the restaurant ahead of the Russians in line by displaying their diplomatic cards. Because of the enormous disparity in the unofficial exchange rate between the US dollar and the Russian ruble, my son could purchase as many as ten Big Macs for the equivalent of just a few US dollars. Eventually, I had to restrict him from attending McDonald's for health reasons. It is well known that fast foods can contribute to serious health issues if consumed excessively. For instance, daily indulgence can increase the risk of developing diabetes due to the high amount of simple carbohydrates in such meals, which can lead to insulin resistance and type 2 diabetes. However, it is worth noting that McDonald's has taken steps over the years to address the calorie content of its meals. Still, I felt it was crucial to be mindful of my family's dietary choices and avoid making McDonald's a daily habit. I believed moderation was key, and balancing indulgent eating with a healthy diet was essential.

The opening of McDonald's in Moscow was symbolic in many respects. First, it was a significant feat of fast-food diplomacy, if not a masterstroke. It represented the prosperity of American capitalism during the collapse of the Soviet system. For most

Russians, opening the first McDonald's in Moscow provided a glimpse of what dining out could be like beyond the Iron Curtain. It symbolized the end of the Cold War and the beginning of a new era of globalization in Russia. It was seen as a sign of Western culture and capitalism entering the heart of the Soviet Union, representing the crumbling of the Iron Curtain and Russia's embrace of the West.

The opening of McDonald's in Moscow also paved the way for other American fast-food chains to enter the Russian market, shaping the country's culinary landscape. My family and I will never forget how the opening of the first McDonald's in Moscow transformed the lives of so many Russian citizens. That was hardly surprising. The opening of McDonald's in Moscow in 1990 symbolized the Soviet Union's economic and political reforms and the beginning of a new era in Soviet life. It was a visible sign of the "crack in the Iron Curtain" and the end of the Cold War—a symbol of change and of the Soviet Union's opening up to the outside world, as well as a harbinger of the Soviet Union's impending collapse. The opening of McDonald's in Moscow was a prominent example of *glasnost,* Soviet President Mikhail Gorbachev's attempt to open up the Soviet Union. Significantly, for the average Russian customer, visiting McDonald's was an opportunity to enjoy a small pleasure in a country embroiled in economic problems and political turmoil. When all was said and done, the opening of McDonald's in Moscow was an epochal moment in a turbulent era.

CHAPTER THIRTY-SIX

AN AFRICAN AMERICAN STUDENT'S FIRST VISIT TO MOSCOW

> *"If the Soviet Union let another political party come into existence, they would still be a one-party state, because everybody would join the other party."*
> \- **Ronald Reagan,** 40th president of the United States

By the time we were embarking on our incredible journey to the U.S. embassy in Moscow, our two older sons, Steven and Marten, were both students at Benedict College, an HBCU in Columbia, South Carolina. *HBCU* stands for Historically Black Colleges and Universities.

HBCUs are institutions of higher education in the United States that were founded to serve the African American community. These colleges and universities have a rich history and have played a significant role in providing educational opportunities to African American students, particularly during segregation. The institutions offer various academic programs, including undergraduate and graduate degrees, and are known for their strong sense of community and commitment to social justice. HBCUs have produced many notable alumni, including civil rights leaders, politicians, artists, and business leaders. They continue to play an essential role in promoting diversity, equity, and inclusion in higher education and beyond.

The two boys were obviously *"missing in action"* concerning our family's foreign assignment, which was unavoidable, as their education had to take vital precedence. However, at the beginning of the summer break that year, Marten announced that he would visit Moscow with us. We were delighted to approve his trip, but Annie and I were understandably apprehensive about Marten traveling from the United States to the Soviet Union alone. On June 10, 1989, Marten finally flew from Washington, DC, to Frankfurt, Germany, to wait for his follow-on flight to Moscow. Upon boarding the aircraft in Frankfurt, Marten noticed that only four American passengers were traveling to Moscow.

When Marten disembarked from the aircraft at Moscow's Sheremetyevo Airport, he was staggered by the enormous size of the KGB security guard detail at the terminal. Their armed presence with submachine guns did nothing to allay Marten's sense of incredulity.

Feeling somewhat subdued, at least compared to his earlier excitement about the imminent Soviet visit, he slowly approached the Immigration and Passport Control desk, where the KGB border guard thoroughly examined his passport and visa before finally allowing him to enter the Soviet Union.

Several members of the Defense Attaché Office received Marten in the Arrival Hall and transported him to the U.S. Embassy. During the thirty-minute drive to the embassy, my son reported that he was struck by the grandeur of Soviet architecture.

When he arrived at our residence on the embassy compound, Annie, Karen, and Torrey were relieved that our son had safely completed his long journey to the Soviet Union and were ecstatic about receiving him. As an African American college student, Marten found visiting Moscow for the first time a unique and eye-opening experience. While exploring the city, he noticed that Moscow, a largely homogeneous city, had minimal racial diversity compared to what he experienced in the United States. Fortunately, the young man did not encounter catcalls or bullying while on the subway, sightseeing around Moscow, or delivering household goods with his White colleagues to embassy staff residing in the city. Yet, it was inevitable that seeing a Black American would spark curiosity and stares from some local Russian citizens.

During his stay in the Soviet Union, Marten noted the stark cultural differences between Moscow and the United States, particularly regarding language, food, customs, and social norms. He found these cultural differences both fascinating and challenging. On several occasions during his visit, I had to explain to my son the disparities between living in the Soviet Union and living in the United States. The two countries had vastly different economic, social, and political systems. For instance, the Soviet Union operated a centrally planned economy, state-owned enterprises, and limited individual freedoms. People were assigned jobs, and the government controlled the means of production, whereas the United States operated on a market-based economy, with private enterprises and fully guaranteed individual freedoms. In America, people could choose their careers, start businesses, and pursue economic opportunities. In the Soviet Union, there was limited access to consumer goods, long lines for necessities, and a general shortage of food and other essential items.

In contrast, the United States had a high standard of living, with a wide range of consumer goods and easy access to food, housing, and other essential services. The Soviet Union provided free education but with a strong emphasis on ideological indoctrination and limited academic freedom. In the United States, there was also free public education, but the focus was on academic freedom, critical thinking, and individual choice.

Marten also learned that personal freedom in the Soviet Union was severely limited, with strict controls on speech, assembly, and travel. This starkly contrasts with the United States, where individual rights and liberties were protected, including freedom of speech, assembly, and travel. In fact, in the Soviet Union, there was strict control on travel, with limited access to foreign countries and a requirement for government permission to travel abroad. Additionally, the Soviet Union operated a largely state-controlled media, with limited access to independent news sources and a strong emphasis on propaganda. This sharply contrasted with the United States, where there was free and independent media, offering a wide range of news sources, publications, and broadcast outlets. These disparities reflected the fundamental differences between the Soviet Union's socialist system and the United States' capitalist democracy. By visiting, my son learned that while the Soviet Union provided basic economic security and social welfare, it came at the cost of individual freedoms and opportunities.

On the second day of Marten's visit to Moscow, I took him on a sightseeing tour of Red Square. Instead of driving through the city center, I opted for a ride on the Moscow Metro, the equivalent of the New York Subway. I believed Marten would find the excursion more educational that way. Rather than taking the most direct subway route to Red Square, I decided to give Marten a tour of the Moscow Metro system so he could experience being a passenger in a crowded subway car. The most ornately decorated subway station was Komsomolskaya Station on the Red Line. Komsomolskaya was known for its grand architecture, intricate design, and stunning artwork, making it one of Moscow's most beautiful metro stations. The station featured high ceilings, marble columns, elaborate chandeliers, and intricate mosaics depicting scenes from Russian lore and culture. Komsomolskaya Station was designed to showcase the Soviet Union's achievements and artistic prowess. The station's opulent design and rich decorations left a lasting impression on Marten.

When we finally arrived at Red Square, Marten was struck by the iconic architecture of the surrounding buildings, such as the colorful domes of St. Basil's Cathedral and the imposing walls of the Kremlin. He also witnessed how many Russians queued up to enter Lenin's Mausoleum.

We also visited the GUM Department Store, a popular shopping mall opposite the Kremlin. GUM offered various products, including clothing, accessories, household items, electronics, food products, and souvenirs. However, in 1989, the availability of certain imported goods from the West was severely limited due to stringent economic restrictions and trade policies. Next, we visited the Rossiya Hotel, located adjacent to Red Square. In 1989, it was one of the largest hotels in the world and a prominent landmark in the city. Marten was astounded to find a Baskin-Robbins Ice Cream shop inside the Rossiya Hotel. To the dismay of local Russian citizens, this establishment did not accept rubles, and all shoppers had to pay for their purchases in "hard currency." Two scoops of ice cream cost about $3.50 in U.S. currency.

Upon returning to our quarters on the embassy compound, Marten told me he was thrilled about the opportunity to visit Moscow. However, his visit was not just to be a tourist. Before returning to the United States, he intended to find employment at the embassy to earn extra money for his college expenses. Over several days, the embassy personnel administration hired Marten for multiple job slots. For example, he worked as a cashier at the embassy bowling alley, a disc jockey in the Marine Security Guards Club, and a laborer delivering household goods to the Soviet apartments of embassy staff living off the embassy compound.

One of the most memorable moments of Marten's visit to Moscow was his attendance at a concert by American entertainer Paul Simon in Gorky Park on July 15, 1989. Paul Simon's performance was one of the first significant rock concerts in the Soviet Union. The concert occurred during political and social changes in the Soviet Union, with the reforms of *perestroika* initiated by Mikhail Gorbachev. The concert attracted a massive crowd estimated to be around 200,000 people, making it one of the most significant music events in the region at that time. The concert symbolized cultural exchanges between the East and West and broke down barriers between the two sides during the Cold War era.

On the eve of his departure for the United States, Marten's friends at the embassy hosted him at a farewell party. As a memento of his time in the Soviet Union, his friends presented him with a Soviet flag, which they all signed. But to Marten's surprise, upon arriving at Sheremetyevo Airport the following day, he noticed that the Soviet flag had mysteriously disappeared from his luggage. Aside from this strange incident, Marten had no problems clearing customs or immigration and successfully boarded his aircraft for the flight to the United States. During his visit to Moscow, he was fascinated by the city's culture, history, and politics while feeling a sense of being in a very different place from what he was used to in the United States. In his later years, Marten would proudly serve our nation as the Associate Director for Procurement Policy at the Executive Office of the President in the White House. To this day, Marten reminisces fondly about his first visit to Moscow.

CHAPTER THIRTY-SEVEN

NOTABLE AMERICAN VISITORS TO MOSCOW BETWEEN 1988 AND 1990

"The trees in Siberia are miles apart. That is why the dogs are so fast."
- Bob Hope

As an Assistant Air Attaché at the U.S. Embassy in Moscow, I had many fulfilling duty schedules, including gathering and analyzing information on Soviet air defense systems, preparing and submitting reports to the United States government on Soviet air power trends, maintaining relationships with Soviet military officers to facilitate seamless communication between the United States Air Force and the USSR's Air Force, and providing administrative and logistical support to the American Air Attaché, particularly for the U.S. Embassy's defense cooperation efforts in Russia. A rather exciting aspect of my job was assisting in planning and coordinating events, such as military-to-military exchanges, visits by U.S. military officials, and visits by prominent American citizens to the Soviet Union. In the process, I met and escorted some notable visitors during my two-year tenure at the U.S. Embassy in Moscow.

Mike Tyson was the world heavyweight boxing champion between 1987 and 1990. Tyson, his then-wife, actress Robin Givens, and his mother-in-law, Ruth Roper Givens, visited Moscow in September 1988. His wife was in Moscow to film two episodes of the television sitcom *'Head of the Class.'* Tyson's visit was a big, if not enthusiastic, celebration. Many Russian citizens came out to catch a glimpse of the famed pugilist. He had the opportunity to meet with fans, visit some of Moscow's famous landmarks, including Red Square, and even spar with local boxers. The Embassy Marine Security Guards adopted "Iron Mike" as their own and often saw him at the Marine Club feasting on hamburgers and chicken wings. Tyson's visit to Moscow was considered a success and helped further cement his status as a global boxing icon.

The Reverend Jesse Jackson, a former U.S. presidential candidate, visited Moscow for the first time in January 1989. He met with Soviet officials to discuss various issues, including human rights, international relations, and arms control. Reverend Jackson's visit was part of his efforts to promote dialogue and diplomacy between the United States and the Soviet Union during a significant regional political change. The U.S. Ambassador invited Annie and me to a formal dinner at his residence in honor of Reverend Jackson's visit to Moscow. During his speech at the dinner, Jesse Jackson commented that, in addition to meeting with Kremlin officials, he planned to meet with Soviet dissidents and seek ways to foster closer communication between American and Soviet citizens at the grassroots level. After leaving Moscow, Jesse Jackson traveled to Armenia to draw attention to the devastating earthquake that had struck the country. His visit helped raise awareness and support for the humanitarian relief efforts in Armenia.

United States Senator Chuck Grassley visited Moscow in August 1989. During his visit, he engaged in diplomatic efforts and discussions about U.S.-Soviet relations. The visit occurred during a significant period of change and transition in the Soviet Union, leading to the eventual dissolution of the Soviet Bloc. At a reception hosted in his honor by the U.S. Ambassador, a congressional staffer approached me and requested that I interpret for the senator and a Russian environmentalist. The thirty-minute discussion focused on the large-scale damage inflicted on people and the environment by the Soviet Union's policy of dumping significant amounts of radioactive waste into the Arctic Ocean. When the senator returned to Washington, D.C., he wrote me a letter expressing his profound gratitude for my assistance as a Russian interpreter. He also stated that he was entering the results of his discussions with the Russian environmentalist into the Congressional Record.

Bob Hope was a British-born American entertainer and comic actor known for his rapid-fire delivery of jokes and one-liners, as well as his success across various entertainment media genres. He was also recognized for his decades of overseas United Service Organizations (USO) tours to entertain American troops, receiving numerous awards and honors for his work as an entertainer and humanitarian. Between 1941 and 1991, he conducted 57 tours for the USO, entertaining military personnel worldwide. In 1997, Congress passed a bill that made him an honorary Armed Forces veteran. I had the privilege of escorting this iconic figure in American entertainment, actress Brooke Shields, and her mother, Teri Shields, when they visited Moscow for the first time in May 1990. The trip was part of a Goodwill Tour to entertain

U.S. military personnel and Embassy employees stationed in the Soviet Union. The visit also aimed to promote friendship and cultural exchange between the two countries during a time of improving relations. On May 17, 1990, approximately 300 spectators, including selected invited Russian guests, watched the performance from the grounds of the U.S. Embassy. The show was a rousing success. Before departing Moscow, Mr. Hope and his party visited Red Square and the new McDonald's restaurant, which had opened in January 1990.

My tour of duty as Assistant Air Attaché at the U.S. Embassy in Moscow eventually came to an end in July 1990. Before I departed, the Air Attaché, a Colonel, wrote the following in my final performance evaluation: *"Following the unexpected humanitarian reassignment of his co-worker, Lt. Col. Wallace found himself responsible for most of what had been done by three officers. He responded superbly, providing effective and error-free support to a record number of USAF diplomatic flights into the USSR."* The Defense Attaché, a Brigadier General, commented, *"As one of my best Russian speakers, Lt. Col. Wallace has been at the forefront of our expanding contacts with the Soviet military."*

Although I was offered the opportunity to be assigned as the Air Attaché at a U.S. embassy in another country, at the end of our two-year tour in Moscow, Annie and I decided that it was time to return to the United States. At my farewell dinner, I thanked each Air Force Attaché team member for their outstanding support during our time in Moscow. We departed Moscow for Washington, D.C., on July 15, 1990, bringing to a close what will easily be considered the apogee of my remarkable career as an officer in the United States Air Force.

CHAPTER THIRTY-EIGHT

BACK IN THE UNITED STATES

"It doesn't take a hero to order men into battle. It takes a hero to be one of those men who goes into battle."
- **General Norman Schwarzkopf Jr.,** Commander of United States Central Command, and Commander of "Operation Desert Shield," the Coalition Force in the Gulf War against the Iraqi invasion of Kuwait.

Upon our return to the United States, my most pressing priority was to find suitable housing for our family. I had been reassigned to the Pentagon as a Russian Area Specialist, and we needed to find a house in a residential community within an hour's commute to the Pentagon. After several days of searching, we found a lovely five-bedroom home in the planned community of Montclair in Prince William County, Virginia. Montclair was known for its scenic surroundings, including a 108-acre lake, numerous parks, recreational amenities, and excellent schools.

With the assistance of our realtor, we discovered a comfortable-looking five-bedroom colonial residence for sale. After viewing the property, we immediately signed a purchase agreement for the full sales price, along with a sizable down payment. To our utter dismay, however, our contract was rejected by the seller. According to our realtor, the owner "refused" to sell his property to African Americans. I faced the harsh reality that, despite having meritoriously served my country for two years as Assistant Air Attaché at the US Embassy in Moscow, certain unpleasant aspects of life in the United States remained unchanged. It left a sour taste in my mouth. Although we were thoroughly disappointed at being rejected for our first choice, we ultimately purchased a beautiful four-bedroom colonial residence in another section of Montclair the following week.

We were genuinely excited about our decision to live in Montclair. At that time, Montclair was not as large as some other towns in Prince William County, Virginia. It wasn't even a town at all; it was a collection of a couple of thousand homes that surrounded or were near Lake Montclair. Perhaps the most charming feature of Montclair is the lake. The lake, easily the community's hub, has three private beaches that residents can use throughout the year. Each beach has a dock, a swim dock, and picnic areas. Excellent common areas surround the lake. People often jog or walk on the miles of sidewalks on Waterway Drive, which encircles the lake. One of Northern Virginia's hidden gems is the Prince William Forest National Park, which borders Montclair to the south. Montclair is also a golf course community, with the Montclair Country Club located within the community. Its fairways and creeks crisscross the area, adding to its beauty and tranquility.

After settling into our new home, I went to work at the Pentagon, while Karen and Torrey started attending Prince William County public schools, and Annie began applying for teaching positions in the Prince William County School District. However, to her amazement, the personnel department rejected all her employment applications. Despite Annie's impressive credentials, it seemed they were not currently hiring any African American teachers. To continue her teaching career, Annie finally applied for an elementary school position in Washington, DC, where she had previously worked before coming to Moscow. After the interview, the personnel department immediately offered Annie a position as a third-grade teacher in a crime-ridden neighborhood in Northwest Washington.

Naturally, I was apprehensive about my wife having to commute to her new job in the inner city for two and a half hours five days a week. However, Annie made it clear that she wanted to continue working to help support our family financially. After two years of teaching in Washington, DC, Annie received the Outstanding Teacher Award from the parents and teachers at Davis Elementary School. In July 1993, by God's grace, Annie finally obtained a permanent teaching position in the Prince William County Schools, where she worked for the next eighteen years before retiring.

When I arrived at the Pentagon in July 1990, the Air Force Personnel Administration assigned me to Headquarters, Air Force Intelligence Agency (AFIS) as Chief of the Russian Air Forces Branch because of my recent tour as an Assistant Air Attaché in Moscow. However, my tour of duty with AFIS was short-lived. On August 2, 1990, at about 2 am local time, Iraqi forces invaded Kuwait, Iraq's tiny, oil-rich neighbor. Kuwait's defense forces were rapidly overwhelmed, and those not destroyed retreated to Saudi Arabia. The Emir of Kuwait, his family, and other government leaders fled to Saudi Arabia, and within hours, Kuwait City had been captured, with the Iraqis establishing a provincial government. By annexing Kuwait, Iraq gained control of 20 percent of the world's oil reserves and, for the first time, a substantial coastline on the Persian Gulf. The same day, the United Nations Security Council unanimously denounced the invasion and demanded Iraq's immediate withdrawal from Kuwait. On August 6, the Security Council imposed a worldwide ban on trade with Iraq.

On August 7, 1990, President George H.W. Bush announced and launched *"Operation Desert Shield"* to deter further Iraqi aggression and protect Saudi Arabia. US forces raced to the Persian Gulf. Iraqi dictator Saddam Hussein, meanwhile, built up his occupying army in Kuwait to about 300,000 troops. On November 29, the U.N. Security Council passed a resolution authorizing the use of force against Iraq if it failed to withdraw by January 15, 1991. Hussein refused to withdraw his troops from Kuwait, which he had claimed as a province of Iraq. Approximately 700,000 allied troops, primarily American, gathered in the Middle East to enforce the deadline. At the start of Operation Desert Shield, the Iraqi Air Force had a significant inventory of Soviet/Russian-made aircraft, reflecting Iraq's close military ties to the Soviet Union during the Cold War. Those aircraft formed the backbone of Iraq's air capabilities. Because of my extensive expertise in Soviet/Russian military planes, the Air Force immediately assigned me as the Air Force Team Chief in the Department of Defense's Joint Intelligence Center.

Five months after Iraqi forces invaded and annexed Kuwait, "Operation Desert Storm" began on January 16, 1991. While sitting at my desk at the Pentagon that day, I vividly remember receiving a call from the Executive Officer to the Joint Chiefs of Staff Chairman, General Colin Powell. According to the Executive Officer, General Powell requested an intelligence briefing on the air situation in Iraq before his scheduled Pentagon Press Briefing with Secretary of Defense Dick Cheney. I immediately grabbed my briefing materials and headed for the Chairman's office.

After introducing myself as the Air Force Team Chief on duty, I began to brief the General on Iraq's current air situation. After I concluded my presentation, General Powell thanked me for the information.

"Colonel Wallace, how many Iraqi aircraft are on your map?" General Powell asked, looking at me sternly.

I replied, providing the number of Iraqi aircraft.

"Colonel, you had better ensure your briefing chart is correct. After I finish my briefing, the press corps will count every aircraft on your chart, and you better have the correct number of aircraft, or I will demote you to Airman," the general added in an ominous yet not unfriendly tone.

Of course, I double-checked my briefing chart to confirm that the number was correct. To my utter amazement, however, the General turned to me and announced, *"I would like you to accompany me to the Pentagon Press Briefing today."*

"Yes, sir!" I replied automatically, my tongue almost clinging to the roof of my mouth. It would have been military sacrilege to say "no" to the Chairman of the Joint Chiefs of Staff of the United States of America.

When we arrived at the Pentagon Briefing Room, Pete Williams, the Assistant Secretary of Defense for Public Affairs, and Secretary of Defense Dick Cheney were already on stage.

General Powell and I joined them, and the briefing began. During the briefing, Secretary Cheney and General Powell provided updates on the progress of the coalition operation against Iraq, including information on military strategies, objectives, and developments in the conflict. Looking back at that historic event, I consider my presence on that stage with both the Chairman of the Joint Chiefs of Staff and the US Secretary of Defense for the first press briefing at the start of Operation Desert Storm to be an incredible experience that I will remember for the rest of my life. C-SPAN, the Cable-Satellite Public Affairs Network, and the local Washington, DC news outlets televised portions of the briefing. To my surprise, later that day, I received a call from my mother, who said she was very proud to see her son on television with General Colin Powell at the Pentagon Press Briefing.

Operation Desert Storm officially ended on February 28, 1991. A ceasefire was achieved through diplomatic efforts, and all parties agreed to abide by the terms outlined in United Nations Security Council Resolution 686. After the war, I returned to my previous assignment at the Pentagon as Chief of the Russian Air Forces Branch at Headquarters, Air Force Intelligence Service. My incredible journey continues.

CHAPTER THIRTY-NINE

OPERATIONS PROVIDE HOPE I AND II

"...Lt Col Wallace has done an outstanding job as mission director, representing this office for relief flights to Armenia and Turkmenistan. Lt Col Wallace's outstanding performance resulted in the flawless airlift and distribution of over 150,000 pounds of medical supplies worth over five million dollars."
- **Robert K. Wolthuis,** Deputy Assistant Secretary of Defense for Global Affairs, Washington, DC, September 1992.

Before I left Moscow in July of 1990, I observed even more unmistakable signs that the dissolution of the Soviet Union was imminent. The prevailing economic stagnation, political turmoil, and internal unrest across all fifteen Soviet republics were potent signals pointing in only one direction. Under Mikhail Gorbachev's leadership, the Soviet Union implemented glasnost, meaning *openness*, and perestroika, meaning *restructuring*, all in a valiant effort to revive the nation's ailing economy and to save a political system seemingly in a war of attrition with itself.

Unfortunately, contrary to original intentions, these reforms merely resulted in greater political turmoil and demands for independence from the Soviet republics, ultimately contributing to the eventual collapse of the Soviet Union. The formal dissolution of the Soviet Union took place on December 26, 1991.

A series of events from 1989 to 1991 led to the final collapse of the Union of Soviet Socialist Republics (USSR), paving the way for the establishment of new, independent republics in the Baltics and Central Asia and creating the Russian Federation.

Formally established in 1922, at its height, the USSR was composed of fifteen republics, the largest of which was Russia. In 1987, seventy years after the 1917 revolution that established a communist state in Russia, the Soviet Union began to decline. Severe economic problems, combined with an international arms race and evidence of imperial overstretch, made reforms necessary for the unified nation to survive. Taking office as chief of the Communist Party of the Soviet Union and the leader of the USSR in 1985, Mikhail Gorbachev attempted to implement the necessary reforms through his dual policies of *perestroika* and *glasnost*, as I mentioned earlier.

Simultaneously, he applied his 'new thinking' to Soviet foreign policy, hoping to end the Cold War with the United States and alleviate the strains the conflict had placed on the economy.

Gorbachev meant well, and although he implemented extensive economic reforms and introduced limited democratic elections, none of his measures were successful in halting economic decline. Unfortunately, he may have been too committed to his fundamental philosophy of socialism to introduce an actual free market system. As he struggled to find solutions to combat stagnation, Boris Yeltsin was elected President of the Russian

Soviet Federative Socialist Republic in a popular election in the spring of 1990. Yeltsin posed a threat to Gorbachev's centralized power for several reasons. As the leader of the largest republic in the USSR, his opinions seemed to hold greater sway in the Soviet body politic. Additionally, because he, unlike Gorbachev, sought credibility by subscribing to a genuine democracy, he achieved a clear victory in an open election, granting his leadership greater legitimacy in the eyes of the people.

Yeltsin also began to propose detaching his republic from the broader Soviet system, suggesting that Russia be granted economic autonomy, and he decided to leave the Communist Party. Throughout the summer of 1990, all fifteen republics of the former Soviet Union followed Russia's lead, declaring themselves sovereign states one after the other.

Although the U.S. Government hoped for continued political stability under Gorbachev, it also sought the emergence of a free market economy in Eastern Europe. This was unacceptable to Gorbachev, who faced increasing challenges from the free market-oriented Boris Yeltsin. Gorbachev initially attempted military crackdowns on republics declaring independence to hold the USSR together. However, he quickly shifted his strategy and announced incentives, such as increased autonomy and a new All-Union Treaty, to keep the breakaway republics in the union. Finally, in the summer of 1991, Gorbachev agreed with several republics that they would become sovereign but remain loosely federated. Before the final agreement was signed, however, a group of Soviet loyalists attempted a coup to preserve the splintering Soviet Union. The coup fell apart after the military refused to comply. In the aftermath, Gorbachev faced blame for allowing the hardliners behind the coup to reach their relatively high positions of power. He stepped aside, and Yeltsin led Russia into the new post-Soviet era, with the total collapse of the Soviet Union in 1991 taking many in the West by surprise.

In January 1992, while working in my office at the Pentagon, I received a phone call from the secretary to the Air Force Intelligence Support Agency commander. She said the colonel would like to meet with me immediately. Upon arrival at the colonel's office, he informed me that due to my previous experience as an Assistant Air Attaché in Moscow and my more-than-passing fluency in the Russian language, he was temporarily reassigning me to the Office of the Deputy Assistant Secretary of Defense for Humanitarian and Refugee Affairs to provide my expertise as a staff planner and mission director for *Operation Provide Hope I.* Under Operation Provide Hope I, the mission of the Department of Defense Office for Humanitarian and Refugee Affairs was to coordinate humanitarian assistance efforts to support the newly independent states of the former Soviet Union. The Department of Defense's role involved providing aid such as food, medical supplies, and winter clothing to alleviate suffering and help stabilize the region in the aftermath of the Soviet Union's collapse.

To put the situation in proper perspective, the Soviet Union was dissolved into independent republics at the end of 1991, and most of these entities joined a confederation called the *Commonwealth of Independent States.* As they entered their first year without a centralized economy, the former Soviet republics suffered severe shortages of food and medical supplies. Western nations organized relief efforts.

On January 23, 1992, Secretary of State James Baker announced *Operation Provide Hope,* an operation to deliver massive amounts of aid to the CIS. Congress appropriated $100 million for relief for the former Soviet republics. Secretary Baker knew that large quantities of food and medicine were available from stockpiles left after the 1991 Persian Gulf conflict. Richard L. Armitage of the State Department and Dr. Robert Wolthuis of the Defense Department worked together to determine the destinations of the cargo, which would be transported by air to save time. Armitage and Wolthuis aimed for the most relief supplies to reach hospitals, schools, orphanages, community shelters, and senior citizen centers. The Military Airlift Command's Gen. Hansford T. Johnson placed Col. John B. Sams, Jr., the commander of the 60th Airlift Wing, in charge of the first phase of the operation, designated Provide Hope I.

From February to May 1992, under Operation Provide Hope I, the U.S. Air Force provided emergency food, medicine, medical equipment, and essential supplies to the former Soviet Republics following the dissolution of the Soviet Union. Sixty-five USAF C-5 Galaxy and C-141 Starlifter cargo aircraft transported thousands of tons of food, medicine, and critical supplies to twenty-four cities in the ten former Soviet republics of *Armenia, Azerbaijan, Belarus, Kazakhstan, Kyrgyzstan, Uzbekistan, Turkmenistan, Tajikistan, Moldova, and Russia.* In my new position within the Department of Defense, I worked closely with the State Department, the Agency for International Development, and other government agencies to coordinate this massive relief effort. I also served as the mission director for numerous humanitarian relief flights to the former Soviet Republics. Operation Provide Hope I provided critical relief, particularly in remote areas, demonstrating the U.S. commitment to stabilizing the former Soviet Republics.

In September 1992, Robert K. Wolthuis, the Deputy Assistant Secretary of Defense for Global Affairs, wrote the following correspondence to the Assistant Chief of Staff, Intelligence, to highlight my performance during Operation Provide Hope I: *"Based on his excellent earlier work on Operation Provide Hope I and his extensive in-country experience, I handpicked Lieutenant Colonel Lew Wallace for special duty with the Office of the Secretary of Defense in connection with the implementation of Operation Provide Hope II. Lt Col Wallace has done an outstanding job as mission director, representing this office for relief flights to Armenia and Turkmenistan. Lt Col Wallace's performance resulted in the flawless airlift and distribution of over 150,000 pounds of medical supplies worth over five million dollars."*

Operation Provide Hope II began in late 1992 and continued into 1993-1994 as an extension of the initial humanitarian relief effort. Under Operation Provide Hope II, there was a shift from emergency aid, food, and medicine to long-term recovery, including infrastructure support, medical system rehabilitation, and agricultural development. During Operation Provide Hope II, multiple U.S. government agencies collaborated to distribute aid effectively, including USAID, DOD, the State Department, and partnerships with NGOs like the Red Cross and Mercy Corps. I worked jointly with the Department of Defense and the State Department during Operation Provide Hope II. After my special duty assignments ended, Ambassador Richard L. Armitage, Deputy to the Coordinator for U.S. Assistance to the Newly Independent States, wrote the following correspondence to the Assistant Chief of Staff, Intelligence: *"Please convey my sincere thanks and deep appreciation to Lt Col Lewis S. Wallace Jr. for his outstanding performance during Operation Provide Hope. I handpicked Lew for the very demanding job of Team Chief in Yerevan, Armenia. In two weeks, working under extremely stressful conditions, Lew spearheaded the delivery and distribution of over one million pounds of food and medical supplies to over one hundred hospitals throughout Armenia. Lew's fluency in Russian and extensive in-country experience were essential during negotiations with government and church leaders. The excellent support provided by Lt Col Wallace ensured a successful and safe operation."*

Operation Provide Hope II paved the way for future U.S. aid programs and underscored the role of military logistics in global humanitarian efforts. Additionally, the unique operation exemplified a strategic blend of altruism and pragmatism, aiming to prevent regional instability while promoting democratic transitions. It also highlighted the U.S. commitment to global humanitarian leadership in the post-Soviet era.

After participating in Operations Provide Hope I and II, General Colin Powell, the Chairman of the Joint Chiefs of Staff, traveled from the Pentagon to the State Department to present me with a prestigious *Humanitarian Service Award* for my exemplary service. My participation in Operation Provide Hope was one of the proudest accomplishments of my entire Air Force career. I had indeed come a long way from the days of the young Black man whose mother believed in him so much as to unequivocally prophesy uncommon success into his life.

CHAPTER FORTY

BECOMING A FULL COLONEL IN THE USAF

"To become a Colonel in the Air Force is to not only reach a pinnacle of leadership, but to step into a role where you can truly shape the future of our skies, guiding the men and women who defend our nation with unwavering commitment and expertise."
- Author Unknown

In June 1992, after completing my temporary duty assignments with the Department of Defense and the State Department under Operations Provide I and II, I returned to my previous position at the Pentagon as Chief of the Russian Branch at Headquarters, Air Force Intelligence Agency (AFIA). During my absence, the Air Force assigned five new intelligence officers to my branch.

These exceptionally well-trained and highly qualified officers were all designated Russian Area Specialists. They had completed a master's in Russian studies or international relations and the Basic Russian Language Course through the Defense Language Institute.

My primary responsibility was to prepare comprehensive analytical reports and briefings on Russian military and political developments for our Air Force customers. It was common practice for my staff to prepare the intelligence briefings, and I would personally present them to the Air Force high command. One day, a colleague, a senior Air Force Intelligence Colonel, visited my office to offer some friendly advice. He said, *"Colonel Wallace, we know you are an expert in your field, and your briefings to the Air Force Chief of Staff have been outstanding. However, if you are going to be promoted to higher military ranks, you must learn to delegate more of these briefings to your subordinates."* In hindsight, it seemed I was blissfully unaware that I was not giving my staff the opportunity to showcase their expertise and communication skills to senior Air Force staff members. Based on that conversation, I immediately informed my staff that, in the future, they would all have the opportunity to prepare and present briefings to the senior staff and that I would adopt a less intrusive attitude in running the branch. I clarified that I would re-commit myself to freely delegating such responsibilities to them, ensuring proper exposure to enhance their career progression. Unsurprisingly, this turn of events was a source of tremendous joy and fulfillment for all my staff, and over the next six months, they received well-earned kudos for their excellent briefings. The fundamental lesson I gleaned from this incident was that to be an effective leader and manager, one must learn to delegate a portion of one's workload to subordinates. However, my enthusiasm for this change in my branch's responsibilities would be relatively short-lived.

In 1993, the U.S. Air Force (USAF) conducted a Reduction in Force (RIF) as part of a drawdown until 1995. During this time, the Air Force separated about 29,000 workers, of which only 3,800 were involuntary RIFs. A reduction in force (RIF) is a permanent termination of employees due to budget cuts, reorganizations, or lack of funding. It is the traditional method that agencies use to reduce their workforce. The USAF RIF was part of post-Cold War downsizing within the military, and intelligence officers were not immune to these cuts. Despite their sterling credentials, four out of five staff members, all Captains (O-3), received separation orders and had to leave the Air Force. I valiantly tried to appeal this decision to Air Force headquarters, but to no avail. The Air Force had spent thousands of dollars educating and training each of these exceptional Russian Area Specialists, and in my opinion, to prematurely separate them from the service at the midpoint of their careers was nothing short of a monumental tragedy.

After flying to various parts of the former Soviet Union over the past year, I was glad to return home and spend quality time with Annie and the kids. An oft-quoted phrase in the military is, *"If the military wanted you to have a family, they would have issued you one."* This sarcastic saying highlights the challenges of balancing military service with family life, implying that the demands of military duty often make it difficult to maintain a stable family unit due to frequent deployments, relocations, and long hours, essentially suggesting that having a family is not always considered ideal for a military career. I must admit that I often found the constant separation from my family while attempting to advance in my Air Force career particularly challenging. As always, Annie remained very supportive and frequently told me she was my *"Number One Cheerleader, no matter what happens on our Air Force journey."*

In 1993, I arrived at the primary promotion zone for advancement to Full Colonel (O-6). Although I had an outstanding record of achievement in the Air Force, I was still apprehensive that I would never be promoted to Colonel. One thing that troubled me was my lack of attendance at any professional military training schools in residence. As a First Lieutenant and Instructor of Russian at the Air Force Academy, my department head had advised me to decline attendance at the Air Force Squadron Officer Course (SOS) in residence due to my then-current teaching responsibilities at the Academy. Overall, attendance at SOS was considered a vital stepping stone in the professional development of Air Force officers, as it prepared them for increased responsibilities and leadership roles within the Air Force. Because I voluntarily declined to attend SOS, I was never selected for any professional military education in residence during the next fifteen years of my Air Force career. However, I completed the Squadron Officer School, Air Command and Staff College, and the National Defense University's National Security Management Program by correspondence. Then, something remarkable happened, seemingly like a bolt out of the blue, regarding my promotion.

In July 1993, I received a call from the Air Force Personnel Office congratulating me on my selection as one of only sixteen Air Force officers chosen to participate in the USAF National Defense Fellowship program (NDFP). The NDFP is a prestigious initiative designed to provide advanced education and professional development opportunities for senior officers and civilians within the USAF. Candidates are selected based on their leadership potential, professional achievements, and commitment to advancing the USAF's mission. The program aims to cultivate strategic thinkers and leaders who can address complex national security challenges and contribute to the long-term defense objectives of the United States.

I was assigned to the Institute for the Study of Conflict, Ideology, and Policy (ISCIP) at Boston University for a one-year post-graduate fellowship. Annie decided to remain in Virginia for the following year to complete her tenure requirements as an elementary school teacher in the Prince William County Public Schools system.

The ISCIP focused on researching and analyzing global conflicts, political ideologies, and public policy issues. Due to my previous position as the Assistant Air Attaché in Moscow, my fluent Russian, and my role as a mission commander for humanitarian relief flights to the former Soviet Republics, I brought to ISCIP a unique spectrum of experience and rare analytical qualities that meshed ideally with the Institute's mission and activities. During my one-year tenure at the Institute, I played a significant role in directing simulation exercises, planning and implementing the *Distinguished Lecture Series,* and managing the important military aspects of the Institute's collective research program. While at Boston University, the Director of the National Security Fellows Program at Harvard University's Kennedy School of Government hand-picked me to interpret for twenty-five Russian general officers, ten of whom were from the Russian Air Force and participants in Harvard's Executive Program. Additionally, I presented a series of lectures on the Russian military and foreign policy at two senior-level seminars on U.S. National Security Affairs at the Naval War College and delivered lectures on Russian military developments at Harvard and the University of Miami.

At the end of my one-year tour of duty as a National Defense Fellow at Boston University in May 1994, Professor Uri Ra'anan, the Institute Director, wrote a personal letter to Dr. Shelia E. Widnall, the Secretary of the Air Force, which read as follows: *"The Institute was most fortunate to be able to avail itself of the capabilities of this fine officer who combines the best traditions of professionalism with qualities of intellect and character that made him ideally suited to be a National Defense Fellow here and to serve as a link with senior personalities in the academic and governmental arenas."* Professor Ra'anan also noted that, given my qualities and experience, he believed I had the potential for senior positions in the Air Force to make the most of my background in Russian military affairs and the Russian language.

I returned to my previous position in Air Force Intelligence at the Pentagon to the delightful news and the official notification of my line number for promotion to the rank of Full Colonel (O-6). To say that I was overjoyed would be an understatement. By the grace of God, through the fervent prayers of my mother and wife, and my persistence and determination, I had finally reached this important milestone in my Air Force career. My incredible journey continues.

CHAPTER FORTY-ONE

ASSIGNMENT TO THE NORTH ATLANTIC TREATY ORGANIZATION (NATO)

"We have a firm commitment to NATO, we are a part of NATO. We have a firm commitment to Europe. We are a part of Europe."
- **Dan Quayle,** 44th Vice President of the United States

The day I received the news about my promotion to the rank of colonel was one of the proudest days of my life. I immediately called Annie to share the good news. To say that she was exhilarated would do a gross injustice to my wife's boundless joy. Annie was thrilled beyond description. Congratulating me with sheer ecstatic fervor, she declared that our hard work and sacrifices during our incredible journey had finally paid off. I couldn't agree with her more.

My next call was to my mother. When I told Mom about my promotion to colonel, she was silent for a moment. Then, she said, *"Son, I told you as a young child that through perseverance, determination, hard work, and a belief in God, you can accomplish all your dreams in life." Soberly, I replied, "Yes, Mom, that was what you said. You were right, but I am proud to have proved you right. Thank you, Mom!"*

In April of 1994, I received a call from the Colonel's Group at the Air Force Personnel Center. The Colonel Group is an elite section that oversees the selection, assignment, and career management of Air Force colonels. The assignments officer told me I had been nominated for a critical Joint Specialty Officer position at the Office of the United States Military Representative to NATO's Military Committee (USMILREP) at NATO Headquarters in Brussels, Belgium. The assignments officer also stated that my nomination for this position must be circulated among all sixteen NATO member countries for review and approval under a so-called *"Silence Procedure."* During this process, member countries have a designated "silence period" to raise any objections or concerns about the proposed appointment. The appointment is approved if no objections are raised within the specified silence period. Fortunately, all NATO member nations approved my assignment based on my outstanding military credentials and recent experience in the USAF National Defense Fellow program. Additionally, in my new position in Brussels, I would again be granted diplomatic status and be accredited as an attaché by the Belgian Ministry of Defense.

The North Atlantic Treaty Organization (NATO) is a military alliance established on April 4, 1949, with the signing of the North Atlantic Treaty in Washington, D.C. The organization aims to provide collective defense against potential security threats, promoting stability and cooperation in the North Atlantic area.

The original twelve founding members of NATO were Belgium, Canada, Denmark, France, Iceland, Italy, Luxembourg, the Netherlands, Norway, Portugal, the United Kingdom, and the United States. Over the years, however, NATO has expanded to include 30 member countries, with the most recent additions being North Macedonia in 2020 and Finland in 2023.

NATO's primary purpose is to provide collective defense against potential security threats, as outlined in Article 5 of the North Atlantic Treaty. NATO's decision-making process is based on consensus, with all member countries having an equal say in the organization's policies and decisions. NATO has a military structure that includes a Supreme Allied Commander (SACEUR) and several subordinate commands, including the Allied Command Operations (ACO) and the Allied Command Transformation (ACT). Interestingly, I was embarking on my NATO assignment at a time of significant activity, with NATO heads of government at the 1994 Brussels Summit committing to a process of enlargement to include democratic states to the East. In fact, NATO welcomed its newest members, the Czech Republic, Hungary, and Poland. This historic step culminated in a process in which President Clinton and other NATO leaders signaled their determination to erase the Cold War division of Europe and strengthen the Alliance by accepting new members.

NATO was, in that era, on the threshold of history. However, the organization was fully committed to its task, as its collective defense commitment is enshrined in Article 5 of the North Atlantic Treaty, which states that an attack on one member state is considered an attack on all member states. Over the years, NATO has been involved in several crisis management operations, including in the Balkans, Afghanistan, and Libya. NATO also engages in cooperative security activities with partner countries, including through the Partnership for Peace (PfP) program and the Mediterranean Dialogue, in addition to providing capacity-building assistance to partner countries through training and education programs.

The United States Military Representative to NATO's Military Committee (USMILREP), a three-star general or admiral, is appointed by the president and confirmed by the Senate. He serves as the principal military advisor to the U.S. Ambassador to NATO and ensures that U.S. military interests are represented in NATO's military decision-making processes. Additionally, the USMILREP represents the Chairman of the Joint Chiefs of Staff and the Secretary of Defense in NATO's Military Committee, which is composed of the Chiefs of Defense of all NATO member nations. The office also coordinates U.S. military positions on NATO policies, operations, and strategic initiatives, ensuring alignment with U.S. national security objectives. As the Air Strategy/Policy Planner on the United States Joint Planning Team, I would assist the USMILREP in policy analysis, coordination, and liaison with NATO counterparts. As a newly selected Colonel, I was utterly amazed that the Air Force had nominated me to NATO for this prestigious assignment. When I called Annie and told her the good news, she was elated. However, when I finally returned home that evening, Annie and I seriously discussed the implications of accepting this new assignment.

The primary challenge was that the tour of duty in Brussels was for three years if she accompanied me or just two years if I went alone without my family. This was a tough decision to make. However, after weighing all the *pros* and *cons,* we decided that Annie would remain in Virginia to continue her teaching career. I would travel to my new assignment alone for two years.

We pledged that during the summer months, when school was not in session, Annie would fly to Brussels so we could spend some "quality" time together. As I learned early in my military career, the needs of the Air Force are always paramount when it comes to new assignments. Now, however, that I was a Colonel Selectee, I called the Colonels Group and asked them to consider the following compromise: *"I would accept the assignment to NATO if they would guarantee that at the end of my two-year tour, I would be reassigned to a position in the Washington, DC area."* Surprisingly, they agreed to my compromise, and I accepted the assignment.

Before departing for my new assignment at NATO as a newly appointed colonel, I requested that the Air Force temporarily promote me to "Full" Colonel before my official promotion date under a process called *"frocking."* Frocked officers are authorized to wear the insignia of the higher rank, such as the silver eagle insignia worn on the uniform collar for colonels. This distinguishes them as officers who hold the rank of colonel even though their official promotion has not been finalized. Due to my high-level position at NATO Headquarters, it was appropriate for me to be frocked as a colonel before arriving in Brussels. This would allow me to transition smoothly into my new roles and immediately begin assuming the responsibilities of a higher rank. Also, upon arriving in Brussels, I discovered I was the first African American colonel to be assigned to this prestigious position at NATO.

During my two-year tour at NATO, I served as the USAF Strategy/Policy Planner on the U.S. Joint Planning Team in the Office of the USMILREP to the NATO Military Committee. My specific duties and accomplishments included:

1. Providing expert advice to the USMILREP and the Chairman of the Joint Chiefs of Staff on all aspects of U.S. and NATO political and military planning.

2. Ensuring that military implications were considered in developing U.S. and NATO positions for Partnership for Peace (PfP) negotiations, Arms Control Negotiations, Nuclear Policy, and Weapons Proliferation issues.

3. Serving as the senior policy officer and country expert for Russia and 25 additional NATO partner nations.

4. Serving as the U.S. representative to the NATO Military Cooperation Working Group, the Military Technical Issues Working Group, the Nuclear Policy Group, and the Weapons Proliferation Panel.

5. Taking a leading role in implementing President Clinton's 100-million-dollar Partnership for Peace Warsaw Initiative, incorporating 27 partner nations with NATO.

6. Being hand-picked by the U.S. Defense Advisor to NATO to escort the Russian Minister of Defense during crucial talks with the Secretary of Defense, facilitating agreements for Russian participation in Operation Joint Endeavor.

While assigned to NATO, I am exceedingly proud of my contributions to the Partnership for Peace (PfP) Program. The PfP program was the NATO initiative designed to foster trust and cooperation between NATO member states and non-member countries, primarily in Europe and the former Soviet Union. Established in 1994, the program built political and military ties, promoted transparency, and enhanced interoperability between NATO and partner nations.

In my final Officer Performance Evaluation before departing NATO, U.S. Army Lieutenant General Thomas M. Montgomery, the USMILREP to the NATO Military Committee, wrote, *"Colonel Lew Wallace remains indispensable as my top political-military advisor. He was commended for his expert participation in a White House-sponsored symposium on European security. At a significant bilateral conference, the National Defense University President lauded Colonel Wallace's briefing on U.S.-Russia relations. Continue challenging this proven performer and leader with the most demanding senior leadership positions, a natural for the National Security Council Staff, then send him back to Moscow as our Defense Attaché."*

In July of 1996, I departed Brussels for my new assignment in Washington, DC, as the Deputy to the President of the Joint Military Intelligence College. After two long years at NATO Headquarters, reuniting with my family and friends was a total and undiluted joy. However, I consider my duty tour at NATO one of the most brilliant highlights of my eventful Air Force career. I am glad that NATO remains vital to European and transatlantic security, providing a framework for collective defense and cooperative security activities. While the organization faces various challenges, including the ongoing conflict in Ukraine and the threat of terrorism, it continues to adapt and evolve to meet the changing security landscape. I remain proud to have been privileged to walk the corridors of the headquarters of that esteemed organization in Brussels.

CHAPTER FORTY-TWO

THE JOINT MILITARY INTELLIGENCE COLLEGE

"The Joint Military Intelligence College has the privilege and trust of serving as the Federal Government's center for intelligence education and research."
- A. Denis Clift, President, Joint Military Intelligence College (1994-2009)

Now known as the National Intelligence University (NIU), the *Joint Military Intelligence College*, also previously known as the Defense Intelligence School and Defense Intelligence College, was established by the United States Department of Defense within the Defense Intelligence Agency in 1962 to consolidate existing U.S. Army and U.S. Navy academic programs in strategic intelligence. In 1980, the U.S. Congress authorized the school to award the Master of Science in Strategic Intelligence degree. In 1981, the Commission on Higher Education of the Middle States Association of Colleges and Schools accredited the school. That same year, the Department of Defense rechartered the institution as the Defense Intelligence College, placing greater emphasis on its research mission. Since then, the university has added several off-campus programs at the National Security Agency and various regional centers and has encouraged an increase in enrollment from civilian agencies. On campus, it also offers two part-time graduate programs. Students from the intelligence community attend the school, including active duty and reserve military personnel from each service, the Coast Guard, the Department of Defense, and other federal civilian employees.

In 1993, the college was renamed the Joint Military Intelligence College, and by the time it featured in my career landscape in the Air Force, it had already embarked on a new era with a more sharply defined mission. I am proud to disclose that I was a key participant in the 1997 congressional authorization of the college to award a Bachelor of Science in Intelligence (BSI) degree. From July 1996 to July 1998, I was the first African American senior military officer to serve as Deputy to the Joint Military Intelligence College (JMIC) President. During my tenure, the JMIC was a federally chartered and accredited graduate and undergraduate institution under the Defense Intelligence Agency (DIA), providing advanced training in intelligence analysis, collection, and operations for members of the U.S. intelligence community. The college served members from all U.S. military branches and civilian intelligence community employees.

The President of JMIC, a distinguished intelligence professional and Senior Executive Service (SES-4) member, provided the college's overall leadership and strategic direction and liaised with the Department of Defense and other government organizations. Because of his position as President, he received numerous invitations to attend National Day receptions at various foreign embassies in the Washington, DC, area. However, he requested that Annie and I represent him at many events because he resided in Annapolis, Maryland. I thanked him for giving us the rare privilege of being active participants in the diplomatic community once again.

A significant highlight of our participation in the Washington diplomatic community was our attendance at the annual Attaché Ball hosted by the Department of Defense in the elaborate diplomatic reception rooms at the State Department. At this event, Annie and I had the pleasure of personally meeting the Secretary of Defense, William Cohen, and his lovely wife, Janet. Cohen, a Republican cabinet member of note during the Clinton era, had previously served in both the United States House of Representatives and the United States Senate. On December 5, 1996, President Bill Clinton announced Cohen's selection as Secretary of Defense, stating that he was the "right person" to build on the achievements of retiring Secretary William Perry *to secure the bipartisan support America's armed forces must have and deserve."* As Secretary of Defense, Cohen played a prominent role in directing U.S. military actions in Iraq and Kosovo.

As the Deputy to the President and senior military officer at JMIC, my job was to direct the college's day-to-day operations. For example, I administered all aspects of the Postgraduate and Undergraduate Intelligence Programs, which included overseeing a $7.6 million annual budget, 435 full- and part-time students, and over 100 faculty and staff members. I also conducted monthly career sessions with students to ensure all graduates received follow-on assignments that maximized their education and training utilization. Finally, I oversaw the production of over 200 military and civilian performance evaluations, promotion recommendations, and awards.

One of my proudest achievements at JMIC was obtaining congressional authorization to establish the Bachelor of Science in Intelligence (BSI) program in 1997. The BSI was a one-year degree completion program for students who had completed three years or equivalent credits (80 semester hours minimum) of study to earn their undergraduate degree in intelligence. This new program was especially beneficial to enlisted intelligence professionals because, after receiving their undergraduate degrees, they could apply for commissions as officers in their respective services.

In April 1998, I received a call from a senior officer, an assignments officer at the USAF Personnel Center. He said that because of my outstanding service record, the Air Force had nominated me to the Department of Defense to return to the U.S. Embassy in Moscow as the Principal Air Attaché. Naturally, I was flabbergasted at this news. For many years, I had dreamed of returning to Moscow as the Air Attaché, with the potential of being promoted to Brigadier General and then later assuming the position of U.S. Defense Attaché. After thanking the assignments officer for his call, I telephoned Annie to share the good news. Although Annie was elated about the possibility of us returning to Moscow for another tour of duty, she said we had to carefully weigh the *pros* and *cons* of accepting this assignment. She mentioned that it might be too stressful at our age to live and work in Russia for another two or three years. Also, at this point in our lives, she added, we should consider retiring to enjoy our golden years with family and friends.

To make the situation even less appealing, I received troubling news during preliminary consultations with the DIA Attaché Office staff. They informed me that if I returned to Moscow as the Air Attaché, I would be required to live in a leased Russian apartment building in downtown Moscow without any security for my family. Having resided in the Embassy compound with my family during my first tour in Moscow, the thought of living in a rundown Russian apartment complex was patently unacceptable. My response to the DIA Attaché Office staffer was crisp and unequivocal: *"I do not want to return to Moscow and live like a Russian, with no security for my family!"* Perhaps my heated response sealed my fate.

About two weeks after this initial meeting, I received a call from the DIA Deputy Director for Attaché Affairs to attend a meeting with him and the DIA Director. At this meeting, the DIA Director informed me that the DIA did not approve my nomination to become the Air Attaché in Moscow. When I asked for more information about the rejection of my nomination, the Deputy Director for Attaché Affairs responded that during my previous tour in Moscow, my wife had constantly telephoned the DIA Attaché Office in Washington, complaining about various problems at the U.S. Embassy in Moscow. That assertion was categorically false. It was nothing short of a trumped-up charge to cast my wife in the worst possible light, seeking the slightest excuse, no matter how flimsy, to discredit my nomination as Principal Air Attaché. Annie never called DIA headquarters to complain about anything.

Next, the general showed me a copy of an article published in "Jet Magazine" in August 1990, describing my experiences as the first Black Air Force Attaché assigned to the U.S. Embassy in Moscow. In the interview, I was quoted as saying, *"Although dramatic change has occurred socially in the U.S.S.R., there is very little change militarily. We will never be allies."* Although my prediction would prove to be an ongoing and fulfilled prophecy, I was personally chastised for conveying this assessment in a public forum without advanced approval from the U.S. government. In other words, as a military officer and former attaché, my comments were considered an enormous diplomatic blunder in some circles in Washington. After our meeting, I thanked the generals for explaining why they rejected me for the Air Attaché post in Moscow.

While I was understandably disappointed and irritated about this unfortunate turn of events, I remembered that my mother always prophetically quoted the scripture in Isaiah 54:17, *"No weapon that is fashioned against you shall prosper."* Two weeks after this contentious meeting with the DIA senior leadership, I received another call from the Colonel's Assignments Group. They informed me that I had been approved for a new assignment as Commander of the Air Force Reserve Officer Training Corps (AFROTC) Detachment 130 at Howard University in Washington, DC.

CHAPTER FORTY-THREE

ASSIGNMENT AT HOWARD UNIVERSITY

"The Mission Command of USAF is a philosophy of leadership that empowers Airmen to operate in uncertain, complex, and rapidly changing environments through trust, shared awareness, and understanding of the commander's intent."
- Mission Command of the Air Force of the United States of America

One of the highest pinnacles of an Air Force officer's career is being selected for a "Command" position. The selection of an Air Force unit commander is a rigorous process that considers a candidate's leadership experience, performance, qualifications, and potential for command. Factors considered during the selection process include the officer's professional competence, character, ability to lead and inspire others, and overall suitability for command. The process is designed to identify the most qualified and capable individuals to lead Air Force units effectively and accomplish missions. My new assignment as Commander, Air Force Reserve Officer Training Corps (AFROTC), Detachment

130, at Howard University in Washington, DC, marked the first time I would have the honor of serving as a commander during my over thirty-year sojourn in the military.

The drive to the grand entrance of Howard University was incredibly nostalgic for me. The neighborhood around Howard University, known as LeDroit Park, has faced challenges with crime and drugs over the past years. I recalled the many days of my youth when I was a member of the notorious LeDroit Park street gang. Now, some thirty years later, here I was, a heavily decorated Air Force Colonel and commander of one of the premier AFROTC units in the United States. I felt a tremendous sense of awe, even as I silently acknowledged the grace and might of God in working remarkable miracles in the lives of seemingly the most unlikely individuals.

Howard University is a private, historically Black, federally chartered research university in Washington, DC. It is classified as "R1: Doctoral Universities – Very high research activity" and is accredited by the Middle States Commission on Higher Education. Established in 1867, Howard offers undergraduate, graduate, and professional degrees in more than 120 programs. Shortly after the American Civil War ended, members of the First Congregational Society of Washington considered establishing a theological seminary to educate Black clergy members. Within a few weeks, the project expanded to include the establishment of a university. Within two years, the university comprised the colleges of liberal arts and medicine. The new institution was named for General Oliver Otis Howard, a Civil War hero who was both the founder of the university and, at the time, the commissioner of the Freedmen's Bureau. Howard later served as president of the university from 1869 to 1874. The U.S. Congress chartered

Howard on March 2, 1867, and much of its early funding came from endowment, private benefaction, and tuition.

The AFROTC program at Howard University is designed to prepare students to become officers in the Air Force. Detachment 130 is a consortium that includes several other universities in the Washington, DC area, such as American University, Catholic University, George Washington University, Georgetown University, Marymount University, Trinity University, and the University of the District of Columbia. Therefore, students from other universities must come to Howard to participate in all AFROTC programs. As Commander of Detachment 130, I played a significant role in overseeing the training, development, and administration of cadets in the program. Some of my specific duties and responsibilities included:

1. Serving as the cadets' primary leader and mentor, guiding their professional and personal development.

2. Overseeing the administration and execution of the AFROTC program at Howard University and other consortium schools.

3. Monitoring cadet performance in academics, physical fitness, and leadership.

4. Directing efforts to recruit new cadets from Howard University and other consortium schools. I promoted the AFROTC program through outreach events and partnerships with academic departments and community organizations.

5. Representing the Air Force and AFROTC at university events, community functions, and military ceremonies.

6. Providing cadets with career advice and mentorship as they transition into active duty.

7. Supervising the detachment's staff, including active-duty Air Force personnel, instructors, and support staff.

8. Ensuring the detachment complies with all Air Force and Department of Defense regulations.

9. Staying current on Air Force policies, leadership practices, and military trends to provide the best possible guidance to cadets.

Overall, I was exceedingly proud to serve as the commander of Detachment 130. It allowed me to shape future Air Force leaders by ensuring that cadets were well-prepared to serve as officers while upholding the Air Force's core values of *"Integrity First, Service Before Self,* and *Excellence in All We Do."*

When I arrived at the offices of the AFROTC detachment in the basement of the Military Science building on Howard's campus, I felt a great sense of joy and enthusiasm for the journey I was about to undertake. However, as soon as I read the official detachment letter posted on the bulletin board in front of the office, my exuberance quickly evaporated. I carefully read the letter and noted numerous misspelled words and grammatical errors. Such mistakes were inconsistent with the Air Force Core Values of *"Excellence in All We Do."* After entering the office and introducing myself to my staff, which included three captains, two senior enlisted Non-Commissioned Officers, and a secretary, I immediately went to my office to contemplate what steps to take to ensure that the detachment operated efficiently and effectively with a minimum of errors while meeting the Air Force's exacting standards and achieving AFROTC goals and objectives. At the first meeting with my staff, I shared highlights of my Air Force career and told them how I had enlisted in the Air Force during the Vietnam War and had now risen to Full Colonel. I also explained that my management style was not autocratic but transformational. My goal was to inspire and motivate each staff member to reach their full potential by building relationships, fostering teamwork and excellence, and encouraging innovation.

As the Detachment Commander, I devoted considerable time to recruiting and retaining the cadet corps. Regrettably, out of 90 students enrolled in the AFROTC program, only 30 cadets were full-time at Howard. I tried valiantly to reverse the trend but to no avail. Apparently, during the late 1990s, many African American students at Howard University had a negative opinion about serving as officers in the military and graduating from college. In their view, many African Americans had experienced discrimination, harassment, and bias while serving in the military, leading to distrust and skepticism about pursuing a career in the armed forces. Some of the most gratifying aspects of my job at Howard were interviewing, recruiting, and funding scholarships for promising and well-qualified candidates for our AFROTC program. I managed a $500,000 budget to fund scholarships, subsistence recruiting, and operations for Howard University and the seven other metro Washington-based universities and colleges. Additionally, my focus and command presence led the detachment to receive an overall "Excellent" rating during the 2000 Headquarters Air Education and Training Command Operational Readiness Inspection. I also increased program enrollment by 74 percent through a renewed emphasis on grassroots recruiting to attract quality cadets. Recognizing the need for additional leadership examples, I developed a senior officer speaker program highlighted by a visit from the Vice Chief of Staff of the Air Force.

In addition to my position as the Detachment 130 Commander, I also held the academic rank of Professor and Head of the

Department of Aerospace Studies under the College of Arts and Sciences. In this capacity, I frequently attended faculty meetings with my university colleagues. I recall that during my first faculty meeting, while wearing my Air Force uniform, several faculty members aggressively approached me and called me a *"Baby Killer"* because I had participated in the Vietnam War. It was evident that some faculty members had a negative impression of the military, and these individuals had probably left the United States for overseas locations to avoid the draft.

Although my initial tour of duty as the Detachment 130 Commander was scheduled for three years, after one and a half years on the job, I decided to contact the Air Force Personnel Center to submit my retirement papers. Of course, I discussed this significant decision with Annie before making it. She merely smiled and said she was overjoyed that we could finally spend our golden years together with family and friends. It is a routine Air Force practice to award a senior colonel or general officer the *Legion of Merit Award upon* retirement to recognize their distinguished and meritorious service during their military career. In the order of precedence, the Legion of Merit ranks sixth below the Medal of Honor. My administrative officer drafted a citation for my Legion of Merit award to the Commander of Headquarters AFROTC.

However, to my utter surprise and dismay, the Headquarters Commander, a USAF Brigadier General, disapproved of my Legion of Merit Award and downgraded my retirement award to the second award of the Meritorious Service Medal. Ignoring my honorable and distinguished service during my thirty-five-year career in the Air Force, the general decided that since I failed to complete my entire three-year duty at Howard University, I did not deserve the Legion of Merit award. Interestingly, in two previous assignments, one to NATO and the other at the Joint Military Intelligence College, the Department of Defense and the Air Force had presented me with two prestigious Defense Superior Service Medals, the fifth military award below the Medal of Honor.

Disgusted with this turn of events, I turned to my faith in God and remembered my mother's soothing words, who always told me never to give up in the face of adversity. I took my concerns directly to the Chief of Staff of the Air Force, General Michael E. Ryan. When I entered the general's outer office without an appointment, the executive officer asked for the purpose of my visit. I responded that I was scheduled to retire with thirty-five years of service from the Air Force. However, my higher headquarters had recently downgraded my Legion of Merit Award to the second award of the Defense Meritorious Service Medal. After speaking with the Chief of Staff's executive officer, he motioned me into the general's office. I introduced myself and handed the general a copy of my Air Force resume. After reading it, the general, with a look of amazement, said, "*Don't worry, Colonel, I will take care of this.*" I thanked the general for his kind words and left his office. Several weeks later, I received a call from AFROTC Headquarters chastising me for not following the established Air Force "chain of command" and for taking my concerns directly to the Air Force Chief of Staff.

On October 31, 2000, I officially retired from the Air Force after thirty-five years of service.

Annie and I planned a monumental retirement service and reception at the Bolling Air Force Base Officers' Club, which we had visited on our first date before getting married. Over 100 family, friends, and former colleagues attended the event. Of course, my mother was exceedingly proud of what I had accomplished over the past thirty-five years in the Air Force. Regrettably, I did not receive my Legion of Merit at my retirement ceremony due to bureaucratic delays. However, I eventually received the medal in the mail about a month after my retirement.

EPILOGUE

MY VALEDICTION

"Nobody starts off as a hero; that does not happen even in films. It is passion, hard work, and perseverance that make the difference. Dream big, follow your passion, and work hard toward fulfilling it, and it will pay off."
- Rakshit Shetty

If you are reading these words, it means you have taken the time to journey with me through the pages of this remarkable biography, *"WINGS OF DIPLOMACY - My Lifetime of Service As An Air Force Attaché."* Please allow me to express my sincere gratitude for making my effort worthwhile. It has been a long and challenging road, yet it has been a valuable and fulfilling one. Writing a book demands a significant amount of time and tremendous effort, requiring near-superhuman discipline. Most importantly, it necessitates commitment and energy at an uncommon level.

In this combined *autobiography* and *memoir*, the narrative chronicles the key events of my incredible career in the United States Air Force, my challenges, and my key achievements. Naturally, such an excursion can also be quite subjective, which is necessary because the entire narrative is based on one's recollection of specific situations and events. Despite that potential handicap,

I have tried to remain within the boundaries of fact, especially as they relate to my primary constituency, the United States Air Force.

I wrote this book to inspire, inform, and enrich readers. My primary motivation for sharing my biography arises from a desire to instill hope in the disadvantaged Black youth of America. That message of hope is that there is a viable alternative to a life of gangs, drugs, crime, and violence. Often, these youth have little or no power over their circumstances. Unfortunately, they find themselves in neighborhoods where they seem doomed to a vicious cycle of helplessness and powerlessness. I am a fortunate survivor of that setting and background, having grown up in abject poverty in the inner city of segregated Washington, DC. Yet, I dreamed of a better life for myself, aiming to escape the shackles of poverty and become a productive member of society. That is why I have written this biography in the fervent hope that it will inspire young individuals from underserved and underprivileged communities. It is also my hope that, within the pages of this book, they will find a blueprint that will help them challenge the stereotype of how difficult it is for a person of color to transcend racial prejudice and discrimination, break free of institutional barriers, and ascend to any height in public service.

I drew immense strength from my mother throughout my distinguished military career, constantly seeking comfort, refuge, and encouragement in her three powerful pronouncements. The first was, *"Son, you can be anything you want in this great country. It just takes believing in yourself, getting a good education, working hard, and never giving up on your dreams."* The second consisted of the words of scripture in Isaiah 54:17, *"No weapon that is fashioned against you shall prosper."* The third was the admonition, *"Son, never give up in the face of adversity."* Every child needs a hero. My mother was my hero. Her words gave me strength and sustained hope through the early years of domestic upheaval and the often heart-wrenching experiences of prejudice I encountered during my military career.

Needless to say, I also drew tremendous strength to persevere from the twin pillars of support represented by my two wives at the polar ends of my Air Force career. First, Vivian, my first wife, provided crucial support in the form of herself and the wonderful children she gave me at the nascent stages of my career. Vivian was a young officer's wife who endured the inevitable absences of a budding Air Force career. My second wife, Annie, entered my life at a critical stage—the significant threshold of realizing my dream to become an Assistant Air Attaché in Moscow. Indeed, if *"the third time was the charm,"* then my dear wife, Annie, was the genie who emerged from the 'bottle of charm' to bless me with my wish.

One reason I chose to commit my life journey to paper is that stories like mine underscore the indispensable role of hard work and resilience in pursuing one's aspirations, despite the myriad obstacles and setbacks that may arise along the way. I am sincerely drawn to advocate for the cultivation of intrinsic motivation and unyielding determination as a driving force to overcome challenges rather than relying on external validation or the allure of fame. Fame and fortune have never been my primary motivation. As I candidly share my odyssey, recounting instances of humiliation and rejection, my numerous triumphs over disappointments, and my unwavering commitment to my goals highlight the importance of perseverance and determination in achieving one's aspirations. I have also attempted to emphasize the value of resilience, discipline, and unswerving dedication in pursuing one's goals, hoping to serve as a beacon of inspiration for others.

As I close this book, I hope I have succeeded in proving that when you are *committed* and *dedicated* to your life's cause, you will eventually succeed. Yet, commitment and dedication are not easy to cultivate and sustain. They require constant effort and discipline, and those who demonstrate commitment and dedication in pursuing their mission remind us of what we can achieve and overcome when we, too, are committed and dedicated to our endeavors. That means *commitment* and *dedication* are not just words but attitudes and actions that can make a difference in our lives.

With this book, I choose to speak about *ambition, purpose,* and *potential in* the lives of the Black youth of America. In this book, you will follow the life trajectory of a poor Black boy from the inner city of Washington, DC, to the hallowed and well-appointed reception rooms of the State Department, to the tradition-rich hallways of the Pentagon, and the diplomatic corridors of America's embassy in Moscow. This is the remarkable saga of an accomplished and acclaimed expert on the Russian language and affairs, a colonel in the United States Air Force, and the first Black Assistant Air Attaché in United States history to be assigned to the American Embassy in Moscow.

As I complete the narration of my incredible journey and my transformation from an inner-city gang member to an Air Force Attaché, Diplomat, Russian Specialist, and Senior Air Force Colonel, I recall the profound words of my esteemed mentor and friend, the late General Colin Powell, 12th Chairman of the Joint Chiefs of Staff, and 65th United States Secretary of State.

He said, *"A dream doesn't become reality through magic; it takes sweat, determination, and hard work."* The great general was right. I hope my life story will propel you to dream big dreams and sincerely commit to realizing them. When all is said and done, to be successful in life, all it takes is unwavering determination, resolute persistence, and stoic perseverance in your quest to achieve your dreams.

Colonel Lewis S. Wallace Jr., USAF (Retired)

Spotsylvania, Virginia,

United States of America

April 2025

PHOTO GALLERY

My Parents, Lewis S. Wallace Sr. and Beverly Gambrell Wallace.

Lewis S. Wallace Jr. Five Years Old.

My Mother. Beverly Gambrell Wallace.

My Grandmother. Maryana Cross Gambrell.

Lewis S. Wallace Jr. McKinley Tech High School.

Airman First Class Lewis S. Wallace Jr.

Our Family. Steven, Karen, Vivian, and Marten.

Teaching Russian at the Air Force Academy.

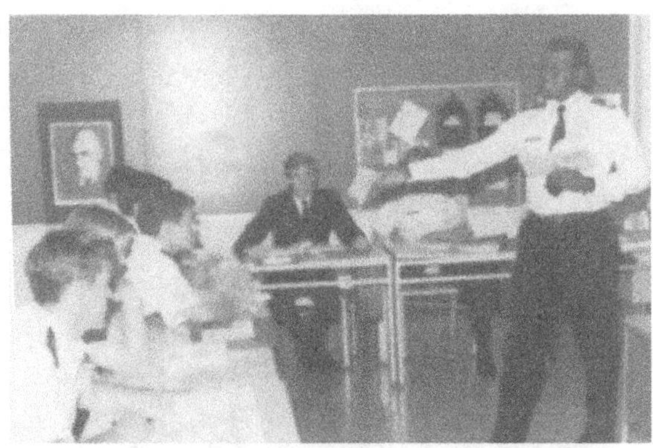

Wedding Photo. Captain and Mrs. Annie Wallace. 1983.

Photo of President Reagan and Mrs. Reagan.

Hosting Attache Dinner. American Embassy, Moscow.

US and Soviet Military Exchange, 1988.

Comedian Bob Hope at American Embassy, Moscow.

Dinner for Rev. Jesse Jackson at Moscow Embassy.

Operation Provide Hope 1993.

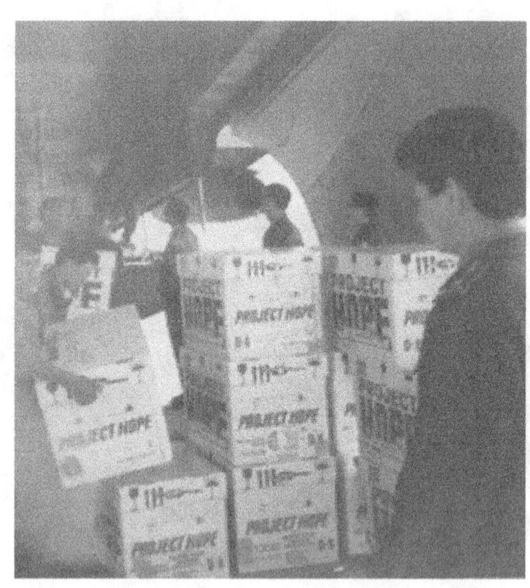

CJCS General Colin Powell with Lt. Col Lewis Wallace.

Former SECDEF and Mrs. Cohen at Attache Reception

Defense Attache Reception. Department of State.

Colonel Wallace with Secretary of State Condoleezza Rice.

Colonel Wallace with Secretary of State John Kerry.

Commander AFROTC, Howard University.

www.ingramcontent.com/pod-product-compliance
Lightning Source LLC
LaVergne TN
LVHW020704110826
845149LV00012B/2104

* 9 7 8 1 9 6 5 5 9 3 7 2 1 *